FUNDAMENTALS OF
MOLECULAR MYCOLOGY

Innovations in Biotechnology

FUNDAMENTALS OF MOLECULAR MYCOLOGY

Devarajan Thangadurai, PhD
Jeyabalan Sangeetha, PhD
Muniswamy David, PhD

AAP | APPLE
ACADEMIC
PRESS

INNOVATIONS IN BIOTECHNOLOGY BOOK SERIES

Devarajan Thangadurai, PhD
Assistant Professor, Karnatak University, Dharwad, South India

BOOKS IN THE SERIES

Volume 1: Fundamentals of Molecular Mycology
Devarajan Thangadurai, PhD, Jeyabalan Sangeetha, PhD, and
Muniswamy David, PhD

Apple Academic Press Inc. | Apple Academic Press Inc.
3333 Mistwell Crescent | 9 Spinnaker Way
Oakville, ON L6L 0A2 | Waretown, NJ 08758
Canada | USA

©2016 by Apple Academic Press, Inc.

First issued in paperback 2021

Exclusive worldwide distribution by CRC Press, a member of Taylor & Francis Group
No claim to original U.S. Government works

ISBN 13: 978-1-77463-577-3 (pbk)
ISBN 13: 978-1-77188-253-8 (hbk)

Library and Archives Canada Cataloguing in Publication

Thangadurai, D., author
Fundamentals of molecular mycology / Devarajan Thangadurai, PhD,
Jeyabalan Sangeetha, PhD, Muniswamy David, PhD.

(Innovations in biotechnology book series ; volume 1)
Includes bibliographical references and index.
Issued in print and electronic formats.
ISBN 978-1-77188-253-8 (hardcover).--ISBN 978-1-77188-254-5 (pdf)
1. Fungal molecular biology. 2. Fungi--Genetics. I. Sangeetha, Jeyabalan, author II. David, Muniswamy, author III. Title. IV. Series: Innovations in biotechnology book series; v. 1

QK604.2.M64T53 2016   572.8295   C2016-900392-2   C2016-900393-0

Library of Congress Cataloging-in-Publication Data

Names: Thangadurai, D. | Sangeetha, Jeyabalan. | David, Muniswamy.
Title: Fundamentals of molecular mycology / Devarajan Thangadurai, PhD,
Jeyabalan Sangeetha, PhD, Muniswamy David, PhD.
Description: Toronto : Apple Academic Press, 2016. | Series: Innovations in biotechnology | Includes bibliographical references and index.
Identifiers: LCCN 2016002302 | ISBN 9781771882538 (hardcover : alk. paper)
Subjects: LCSH: Fungal molecular biology. | Mycology. | Phytopathogenic fungi. | Fungal diseases of plants--Molecular aspects. | Fungi--Genetics.
Classification: LCC QK604.2.M64 T43 2016 | DDC 579.5--dc23
LC record available at http://lccn.loc.gov/2016002302

Apple Academic Press also publishes its books in a variety of electronic formats. Some content that appears in print may not be available in electronic format. For information about Apple Academic Press products, visit our website at **www.appleacademicpress.com** and the CRC Press website at **www.crcpress.com**

CONTENTS

LIST OF ABBREVIATIONS

ABTS	2,2′-azino-bis-3-ethylbenzthiazoline-6-sulphonic acid
AFLP	Amplified fragment length polymorphism
AFTOL	Assembling the fungal barcode of life
AMFEP	Association of Manufactures and Formulators of Enzyme Products
AMT	Agrobacterium mediated transformation
APNEDP	Atmospheric pressure non-equilibrium discharge plasma
AP-PCR	Arbitrary primed polymerase chain reaction
ASAP	Allele-specific associated primer
BCAs	Biological control agents
BFCs	Biofuel cells
BLASTP	Basic local alignment search tool for proteins
BPA	Bisphenol-A
CAZymes	Carbohydrate-active enzymes
cDNA	Complimentary DNA
CGD	Candida genome database
COX1	Cytochrome c oxidase 1
CYGD	Comprehensive yeast genome database
DAF	DNA amplification fingerprinting
DBP	Di-n-butyl phthalate
DDBJ	DNA Data Bank of Japan
DEHP	Di-2-ethyl hexyl phthalate
DES	Diethylsulfate
EMBL	European Molecular Biology Laboratory
EMS	Ethyl methane sulfonate
ESTs	Expressed sequence tags
ETS	Externally transcribed spacer
FCPD	Fungal cytochrome P450 database
FGDB	*Fusarium graminearum* genome database
FPase	Filter paper cellulase

FTFD	Fungal transcription factor database
FunSecKB	Fungal secretome knowledgebase
FUNYBASE	FUNgal phYlogenomic dataBASE
GM-CSF	Granulocyte macrophage colony stimulating factor
HCT	Horizontal chromosome transfer
HMMs	Hidden-Markov models
HVRs	Hyper variable regions
IGS	Intergenic spacer
INSD	International Nucleotide Sequence Database
ITS	Internally transcribed spacer
LiP	Lignin peroxidase
Mbp	Million base pair
MMS	Methyl methane sulfonate
MnP	Mn-dependent peroxidase
NAGA	N-acetylglucosamine
NIG	National Institute of Genetics
NP	Nonylphenol
NPR1	Non-expressor of pathogenesis-related genes 1
NTG	N-methyl-N'-nitro-N-nitrosoguanidine
NTS	Non-transcribed spacer
OMW	Olive mill waste
ORFs	Open reading frames
PAGE	Polyacrylamide gel electrophoresis
PCP	Pentachlorophenol
PCR	Polymerase chain reaction
PLB	Phospholipase B
PPO	Polyphenol oxidase
Pr1	Subtilisin-like proteinase
RAPD	Random amplified polymorphic DNA
REs	Restriction endonucleases
RFLP	Restriction fragment length polymorphism
rpb2	Ribosomal polymerase B2
SCAR	Sequence-characterized amplified regions
SCP	Single cell protein
SIMAP	SImilarity MAtrix of Protein Sequence
SMC	Spent mushroom compost

SRAP	Sequence-related amplified polymorphism
SSCP	Single-strand conformation polymorphism
SSRs	Simple sequence repeats
STMS	Sequence tagged microsatellite markers
STRs	Short tandem repeats
TAPs	Transcription-associated proteins
tef1	Translation elongation factor 1 gene
TFM	Transcription factor matrices
TFM-Explorer	Transcription factor matrix explorer
TNT	2,4,6-trinitrotoluene
VNTRs	Variable number tandem repeats

PREFACE

Mycology is the branch of biology that deals with the study of fungi. Fungi are the diverse group of microorganisms and playing a vital role in the commercial application in agriculture, food, environment and medicine due to its simple model systems. Many fungi produce biological active compounds as secondary metabolites, which have potential to use in various beneficial applications based on its characteristics. Molecular mycology is playing an important role in fungal applications. Also, advances in molecular tools have created a new path for the mycological research and applications in different sectors. Molecular mycology includes molecular markers, DNA techniques, cloning and bioinformatics. This book covers most important aspects in molecular mycology and focuses on the application of fungal secondary metabolites in ecosystem management and sustainable agriculture, application of DNA recombinant techniques to improve industrially important fungal species, different molecular markers and genetic approaches for the taxonomical identification of fungi and also dealt with the bioinformatics tool for the identification fungal species and its secondary metabolites.

This book provides a complete overview of recent developments and applications in molecular mycology. It serves as a comprehensive guide for the identification of fungi and the application of fungal biomolecules in agriculture, food, environment, and pharmaceutical sectors by providing detailed information about application of molecular markers and bioinformatics tools for mycology.

This book has been well written and is based on the authors' renowned expertise and contribution to the various aspects of fungal biology. This book is an excellent source for molecular mycology tools and applications in various fields. This book is equally very suitable for undergraduate and postgraduate biology students and biotechnologists from research institutes, academia and for industry. The authors are indebted to Mr. Ashish Kumar, Apple Academic Press, and AAP staff members for their foresight and valuable and diligent support throughout the publishing task.

—Devarajan Thangadurai, PhD
Jeyabalan Sangeetha, PhD
Muniswamy David, PhD

ABOUT THE AUTHORS

Devarajan Thangadurai, PhD

Devarajan Thangadurai, PhD, is Assistant Professor at Karnatak University in South India, President of the Society for Applied Biotechnology, and General Secretary for the Association for the Advancement of Biodiversity Science. In addition, Dr. Thangadurai is Editor-in-Chief of several journals, including *Biotechnology, Bioinformatics and Bioengineering; Acta Biologica Indica; Biodiversity Research International*; and the *Asian Journal of Microbiology*. He received his PhD in Botany from Sri Krishnadevaraya University in South India. During 2002–2004, he worked as CSIR Senior Research Fellow with funding from the Ministry of Science and Technology, Government of India. He served as Postdoctoral Fellow at the University of Madeira, Portugal; University of Delhi, India; and ICAR National Research Centre for Banana, India (2004-2006). He is the recipient of the Best Young Scientist Award with a Gold Medal from Acharya Nagarjuna University (2003) and the VGST-SMYSR Young Scientist Award of the Government of Karnataka, Republic of India (2011). He has edited/authored 15 books including *Genetic Resources and Biotechnology* (3 vols.), *Genes, Genomes and Genomics* (2 vols.), and *Mycorrhizal Biotechnology* with publishers of national and international reputation.

Jeyabalan Sangeetha, PhD

Jeyabalan Sangeetha, PhD, is a UGC Kothari Postdoctoral Fellow at Karnatak University, Dharwad, India. Dr. Sangeetha was the recipient of Tamil Nadu Government Scholarship (between 2004 and 2008) and the Rajiv Gandhi National Fellowship of University Grants Commission from the Government of India for her doctoral studies. She has published several articles on the effect of pollutants on the environment, and she has organized conferences, seminars, workshops, and lectures. Her main research interests are in the areas of environmental microbiology and environmental biotechnology, with particular emphasis on solid waste management,

environmental impact assessment, and microbial degradation of hydrocarbons. Her scientific and community leadership has included serving as
an editor for the journal *Biodiversity Research International* and acting as
Secretary of the Society for Applied Biotechnology. Dr. Sangeetha earned
her BSc in Microbiology (2001) and PhD in Environmental Sciences
(2010) from Bharathidasan University, Tiruchirappalli, Tamil Nadu, India.
She holds also an MSc in Environmental Sciences (2003) from Bharathiar
University, Coimbatore, Tamil Nadu, India.

Muniswamy David, PhD
Muniswamy David obtained his PhD in Zoology from Sri Krishnadevaraya
University, Anantapur, South India. He joined Karnatak University,
Dharwad, South India, in 1995 as Assistant Professor and advanced to
the rank of Full Professor of Zoology in 2012. He has vast teaching and
research experience in the fields of ecotoxicology, molecular toxicology,
microbial remediation of xenobiotics, and fishery biology. He has published more than 80 papers in national and internationally reputed journals. He has awarded a Postdoctoral Fellowship by the University Grant
Commission, New Delhi, India, and also awarded as "Scientist of the Year
– 2013" from the National Environmental Science Academy, New Delhi,
India. He has successfully handled three major research projects funded
by UGC, DST-SERB, and Hutti Gold Mines Pvt. Ltd., Karnataka.

CHAPTER 1

MOLECULAR MYCOLOGY: AN INTRODUCTION

CONTENTS

1.1 KINGDOM FUNGI

The Kingdom Fungi is being studied extensively across the globe due to their beneficial role in diverse fields namely medicine, fermentation, food industry, bioremediation, biofertilizers and biopesticides. It is also true that most of the plant diseases are caused by fungi, resulting in heavy losses in the crop yield. Fungi vary in size from single cell to multi cell and divergent group of fungi is composed of yeasts, molds, mushrooms, lichens, disease causing rusts and smuts. Due to increase in population worldwide and depleting environment, man is trying to find novel advanced techniques to meet the population demands; and this purpose was resolved through the advancements in molecular biology, which is being utilized in every aspect of biological science. Over the past five decades, advancements in the field of molecular biology has opened up a new array of diagnostic tools to the researchers for addressing issues from gene expression to genetic diversity. The ready availability of DNA sequence data has led to major developments in the study of the biochemistry, ecology, strain characterization, identification, pathogen detection, genomics, molecular diagnosis of fungi, and association with other organisms (Bruns et al., 1991). A molecular marker is a short fragment of DNA that is associated with certain gene in the genome or the marker that reveal variations at DNA level. The first DNA based molecular marker developed is restriction fragment length polymorphism (RFLP) by Grodzicker et al. (1974).

Molecular markers have many advantages over traditional biochemical markers, as they provide data for accurate analysis, but also possess unique genetic characteristics that make them more useful than any other genetic markers (Hillis and Dixon, 1991). Molecular markers render multiple distinguishing advantages; DNA can be isolated with no difficulty from any living constituent such as blood cells, tissues, sperm and hair follicle. DNA samples can be stored for extended period and DNA analysis can be implemented at any stage of cell cycle. Once the DNA is extracted, it is transferred on filter membranes and it can be repetitively hybridized with different probes. In addition, heterologous probe and *in vitro* synthesized oligonucleotide probes can be utilized and PCR-based methods can be subjected to automation (Tibayrenc, 2005). Thus, molecular markers are extensively used for studying genetic diversity, gene mapping, linkage analysis and phylogenetic analysis of fungi.

1.2 INDUSTRIAL POTENTIAL OF FUNGAL APPLICATIONS

Fungi are one of the major organisms with many industrial applications because of its enormous production of economically important metabolites. Use of fungi is spanning from brewing and wine making to biofuel production. A large number of fungal strains have been used for producing antibiotics, drugs, recombinant proteins, organic acids and other useful compounds. Its applications can be seen in various industries in different ways. Fermented foods, sauces and drinks, using fungi became inevitable to the life. Besides metabolite production, its short growth period also found to be attractive in the industrial applications. The choice of the strain may depend on the highest levels of activity in a specific strain.

Quantification of the metabolic fluxes using proper analytical means is an important preliminary step in strain improvement.

The production of the desired metabolite can be maximized by optimization of the environment. After optimization, by changing the environmental parameters, stress induced over production can be obtained. These induction parameters include mutagens also. The use of different mutagenic agents for strain improvement was demonstrated by Parekh et al. (2000). Simultaneous treatment with different mutagens such as N-methyl-N'-nitro-N-nitrosoguanidine (NTG), ethidium bromide and UV or NTG combined with ethidium bromide to create mutant fungi that produced more CMCase and filter paper cellulase (FPase) than wild type fungi was reported by Chand et al. (2005). Random mutagenesis and screening was carried out to improve the production of antifungal metabolites and antagonistic potential of biocontrol agents in fungal strains such as *Trichoderma* spp. and *Gliocladium* spp. (Wafaa and Mohamed, 2007).

1.3   FUNGAL METABOLIC ENGINEERING

Metabolic engineering that is achieved by modification of specific biochemical reactions or introduction of new ones with the use of recombinant DNA technology can bring forth definite product formation (Stephanopoulos, 1999; Nielsen, 2001). Reverse (inverse) metabolic engineering can also be employed for the improvement of the strain which involves: (1) choosing a strain which has a favorable cellular phenotype, (2) evaluating and determining genetic and/or environmental factors that confer specific phenotype, and (3) transferring that phenotype to a second strain via direct modifications of the identified genetic and/or environmental factors (Bailey et al., 1996; Santos and Stephanopoulos, 2008).

Selection of the best strains and improvement of those strains using various classical methods, such as mutagenesis and genetic recombination, were found to be with drawbacks like unspecificity, whereas, strain improvement by genetic engineering has been established as a feasible alternative for them (Verdoes et al., 1995). By single step improvement approach of these traditional methods, production levels were found to be only modestly increasing as two fold or even less (Verdoes et al., 1995). Modification in the genetic level of these molecules will consequently be beneficial to all the fields such as pharmaceuticals and agriculture. Genetic engineering which includes strain improvement through molecular and genetic manipulation is found to be more efficient than the classical approaches.

1.4   FUNGAL BIOINFORMATICS

Bioinformatics has emerged as an integral part in many areas of biotechnology (Zauhar, 2001; Rao et al., 2008). Bioinformatics develops software tools for storage, retrieval, organization and analyzing biological data (Baker et al., 1999; Brown and Botstein, 1999; Neufeld and Cornog, 1999; Backofen and Gilbert, 2001; Lander, 2001). Bioinformatics is extensively used in full genome analysis and indexing the data of organisms which have been sequenced from late 1990's, including microorganisms, nematodes, plants, human and even pathogens (Venter, 2001; Heller, 2002; Friedman, 2004; Kane and Jeffrey, 2007; Wayne and Wishart, 2007). Earlier, identification, classification and phylogenetic analysis of an organism were based on painstaking experiments on living cells and organism (Nasir et al., 2014). Statistical analysis of the homologous recombination rates of different genes determined gene order on a specific chromosome (Schmuckli-Maurer et al., 2003). Based on such experiments, data were combined to create a genetic map, which specified the approximate location of known genes relative to each other. But today, with comprehensive genome sequence and powerful computational tools available to the research community, genome analysis has been redefined (Goffeau et al., 1996; Adams et al., 2000; Douglas et al., 2001; Katinka et al., 2001; Lander et al., 2001; McPherson et al., 2001; Aparicio et al., 2002; Carlton et al., 2002; Gardner et al., 2002; Goff et al., 2002; Holt et al., 2002; Wood et al., 2002; Galagan et al., 2003; Armbrust et al., 2004; Martinez et al., 2004). Our molecular techniques currently been used are very robust which aids in sequencing of new genomes at a far faster rate compared to tools present to annotate them (Thangadurai and Sangeetha, 2013). Hence our ability to obtain genomic data is quickly overtaking our ability to annotate genes. It is evident that manual annotation cannot be scaled to meet the input of newly sequenced organism. The biggest challenge faced by molecular biologists today is that to make sense of the huge wealth of data produced by the genome sequencing projects (Lal et al., 2013). Currently more than 100 fungal genomes have been sequenced and present generation sequencing technologies will further accelerate the accumulation of data over the next decade (Hane and Oliver, 2008; Louis, 2011; Kim et al., 2013). The enormous amount of data produced in this genomic era badly needs incorporation of computers in research techniques to describe features of new genomes; unlike traditional methods, where molecular research was entirely carried out at the experimental laboratory bench space. Generation of sequence, its storage, sequence analysis and interpretation are completely computer dependent tasks. Since molecular organization of an organism is very complex, research is simultaneously carried out in different areas including genomics, proteomics, transcriptomics and metabolomics. Due to which, increase in data volume have been observed in these fields. Intelligent and efficient storage of this amount of data is the challenge faced by bioinformatics community. The stored data on its own is meaningless before analysis and because of the huge volume of data present, it's impossible even for a trained biologist to interpret it manually. Hence computational tools must be developed for the extraction of meaningful biological information from the stored data. Bioinformatics tools are developed based on three central biological principles, e.g., DNA sequences determines protein sequence, which in turn determines protein structure; this protein structure determines protein function. The collaboration of the information obtained from these key biological processes has aided in achieving the long-term goal of the complete understanding of the biology of organisms (Wooley and Lin, 2005).

1.5 FUNGAL SYSTEMATICS

Fungi are an ancient group of 3.5 billion years old according to fossil evidence, whereas earliest fungal fossils are from the Ordovician, 460 to 455 million years old (Redecker et al., 2000). The lifecycle, biosystem response and reproduction of fungal molecule itself can prove their enormous potential in ecosystem and agriculture. Apart from symbiotic relation with ants and mutualism with termites, eukaryotic spore producing achlorophyllous organisms with absorptive nutrition, fungi shows firm relationship with plants. Fungi play a major role in ecosystem as well as in agriculture. It is mainly because of its detriment loving nature. Most fungi are associated with plants as saprotrophs and decomposers. The enzymes secreted by them convert the fats, carbohydrates and nitrogen compounds of the dead animals and plants into simpler compounds, for instance carbon dioxide, ammonia, hydrogen sulfide, water and some other nutrients in a form available to green plants. They break down all kinds of organic matter like wood and other types of plant material that is composed primarily of cellulose, hemicellulose and lignin. Fungi are one of the organisms, which can effectively break down wood. The presence of the fungi is very important for stabilizing the ecosystem. The kind of fungal inhabitation purely depends upon the geographical conditions and season. Thereby, occurrence of the fungal molecule in a specific area may not be stable. As both agriculture and ecosystem management is controlled by fungi, a profound knowledge on the fungal community is necessary in eco- and agro-studies. Fungal biotechnology ends with huge industrial application where as it always commences with cellulosic degradation and so oxidative and hydrolytic enzyme production. Mycorrhizal and parasitic communities in different habitats are well characterized at molecular level and they directly affect plant community structure, composition and productivity.

KEYWORDS

- Antifungal metabolites
- Biochemical markers
- Biocontrol agents
- Biofertilizers
- Biofuel production
- Bioinformatics
- Biopesticides
- Bioremediation
- Brewing
- Cellular phenotype

- **Ecosystem management**
- **Ethidium bromide**
- **Fermented foods**
- **Full genome analysis**
- **Fungal biotechnology**
- **Gene expression**
- **Gene mapping**
- **Genetic diversity**
- **Genetic engineering**
- **Genetic manipulation**
- **Genetic map**
- **Genetic recombination**
- **Genome sequencing**
- **Genomics**
- **Homologous recombination**
- **Kingdom Fungi**
- **Linkage analysis**
- **Metabolic engineering**
- **Metabolite production**
- **Metabolomics**
- **Molecular marker**
- **Mutagenesis**
- **N-methyl-N´-nitro-N-nitrosoguanidine**
- **Oligonucleotide probes**
- **Organic acids**
- **Pathogen detection**
- **Phylogenetic analysis**
- **Plant diseases**
- **Proteomics**
- **Random mutagenesis**
- **Recombinant DNA technology**
- **Recombinant proteins**

- Restriction fragment length polymorphism
- Reverse metabolic engineering
- Strain characterization
- Strain improvement
- Transcriptomics
- Wine making

REFERENCES

Adams, M. D., Celniker, S. E., Holt, R. A., Evans, C. A., Gocayne, J. D., Amanatides, P.
  G., Scherer, S. E., Li, P. W., Hoskins, R. A., Galle, R. F., George, R. A., Lewis, S. E.,
  Richards, S., Ashburner, M., Henderson, S. N., Sutton, G. G., Wortman, J. R., Yandell,
  M. D., Zhang, Q., Chen, L. X., Brandon, R. C., Rogers, Y. H., Blazej, R. G., Champe,
  M., Pfeiffer, B. D., Wan, K. H., Doyle, C., Baxter, E. G., Helt, G., Nelson, C. R.,
  Gabor, G. L., Abril, J. F., Agbayani, A., An, H. J., Andrews-Pfannkoch, C., Baldwin,
  D., Ballew, R. M., Basu, A., Baxendale, J., Bayraktaroglu, L., Beasley, E. M.,
  Beeson, K. Y., Benos, P. V., Berman, B. P., Bhandari, D., Bolshakov, S., Borkova, D.,
  Botchan, M. R., Bouck, J., Brokstein, P., Brottier, P., Burtis, K. C., Busam, D. A.,
  Butler, H., Cadieu, E., Center, A., Chandra, I., Cherry, J. M., Cawley, S., Dahlke, C.,
  Davenport, L. B., Davies, P., Pablos, B., Delcher, A., Deng, Z., Mays, A. D., Dew, I.,
  Dietz, S. M., Dodson, K., Doup, L. E., Downes, M., Dugan-Rocha, S., Dunkov, B. C.,
  Dunn, P., Durbin, K. J., Evangelista, C. C., Ferraz, C., Ferriera, S., Fleischmann, W.,
  Fosler, C., Gabrielian, A. E., Garg, N. S., Gelbart, W. M., Glasser, K., Glodek, A.,
  Gong, F., Gorrell, J. H., Gu, Z., Guan, P., Harris, M., Harris, N. L., Harvey, D.,
  Heiman, T. J., Hernandez, J. R., Houck, J., Hostin, D., Houston, K. A., Howland, T.
  J., Wei, M. H., Ibegwam, C., Jalali, M., Kalush, F., Karpen, G. H., Ke, Z., Kennison,
  J. A., Ketchum, K. A., Kimmel, B. E., Kodira, C. D., Kraft, C., Kravitz, S., Kulp, D.,
  Lai, Z., Lasko, P., Lei, Y., Levitsky, A. A., Li, J., Li, Z., Liang, Y., Lin, X., Liu, X.,
  Mattei, B., McIntosh, T. C., McLeod, M. P., McPherson, D., Merkulov, G., Milshina,
  N. V., Mobarry, C., Morris, J., Moshrefi, A., Mount, S. M., Moy, M., Murphy, B.,
  Murphy, L., Muzny, D. M., Nelson, D. L., Nelson, D. R., Nelson, K. A., Nixon, K.,
  Nusskern, D. R., Pacleb, J. M., Palazzolo, M., Pittman, G. S., Pan, S., Pollard, J.,
  Puri, V., Reese, M. G., Reinert, K., Remington, K., Saunders, R. D., Scheeler, F.,
  Shen, H., Shue, B. C., Sidén-Kiamos, I., Simpson, M., Skupski, M. P., Smith, T.,
  Spier, E., Spradling, A. C., Stapleton, M., Strong, R., Sun, E., Svirskas, R., Tector,
  C., Turner, R., Venter, E., Wang, A. H., Wang, X., Wang, Z. Y., Wassarman, D. A.,
  Weinstock, G. M., Weissenbach, J., Williams, S. M., Woodage, T., Worley, K. C.,
  Wu, D., Yang, S., Yao, Q. A., Ye, J., Yeh, R. F., Zaveri, J. S., Zhan, M., Zhang, G.,

Zhao, Q., Zheng, L., Zheng, X. H., Zhong, F. N., Zhong, W., Zhou, X., Zhu, S., Zhu, X., Smith, H. O., Gibbs, R. A., Myers, E. W., Rubin, G. M., Venter, J. C. (2000). The genome sequence of *Drosophila melanogaster*. *Science* 287(5461), 2185–2195.

Aparicio, S., Chapman, J., Stupka, E., Putnam, N., Chia, J. M., Dehal, P., Christoffels, A., Rash, S., Hoon, S., Smit, A., Gelpke, M. D., Roach, J., Oh, T., Ho, I. Y., Wong, M., Detter, C., Verhoef, F., Predki, P., Tay, A., Lucas, S., Richardson, P., Smith, S. F., Clark, M. S., Edwards, Y. J., Doggett, N., Zharkikh, A., Tavtigian, S. V., Pruss, D., Barnstead, M., Evans, C., Baden, H., Powell, J., Glusman, G., Rowen, L., Hood, L., Tan, Y. H., Elgar, G., Hawkins, T., Venkatesh, B., Rokhsar, D., Brenner, S. (2002). Whole-genome shotgun assembly and analysis of the genome of *Fugu rubripes*. *Science* 297(5585), 1301–1310.

Armbrust, E. V., Berges, J. A., Bowler, C., Green, B. R., Martinez, D., Putnam, N. H., Zhou, S., Allen, A. E., Apt, K. E., Bechner, M., Brzezinski, M. A., Chaal, B. K., Chiovitti, A., Davis, A. K., Demarest, M. S., Detter, J. C., Glavina, T., Goodstein, D., Hadi, M. Z., Hellsten, U., Hildebrand, M., Jenkins, B. D., Jurka, J., Kapitonov, V. V., Kröger, N., Lau, W. W., Lane, T. W., Larimer, F. W., Lippmeier, J. C., Lucas, S., Medina, M., Montsant, A., Obornik, M., Parker, M. S., Palenik, B., Pazour, G. J., Richardson, P. M., Rynearson, T. A., Saito, M. A., Schwartz, D. C., Thamatrakoln, K., Valentin, K., Vardi, A., Wilkerson, F. P., Rokhsar, D. S. (2004). The genome of the diatom *Thalassiosira pseudonana*: ecology, evolution, and metabolism. *Science* 306(5693), 79–86.

Backofen, R., Gilbert, D. (2001). Bioinformatics and constraints. *Constraints* 6, 141–156.

Bailey, J. E., Sburlati, A., Hatzimanikatis, V., Lee, K., Renner, W. A., Tsai, P. S. (1996). Inverse metabolic engineering: a strategy for directed genetic engineering of useful phenotypes. *Biotechnol. Bioeng.* 52, 109–121.

Baker, P. G., Goble, C. A., Bechhofer, S., Paton, N. W., Stevens, R., Brass, A. (1999). An ontology for bioinformatics applications. *Bioinformatics* 15, 510–520.

Brown, P. O., Botstein, D. (1999). Exploring the new world of the genome with DNA microarrays. *Nat. Genet.* 21(S1), 33–37.

Bruns, T. D., White, T. J., Taylor, J. W. (1991). Fungal molecular systematics. *Annu. Rev. Ecol. Syst.* 22, 525–564.

Chand, P., Aruna, A., Maqsood, A. M., Rao, L. V. (2005). Novel mutation method for increased cellulase production. *J. Appl Microbiol.* 98, 318–323.

Douglas, S., Zauner, S., Fraunholz, M., Beaton, M., Penny, S., Deng, L. T., Wu, X., Reith, M., Cavalier-Smith, T., Maier, U. G. (2001). The highly reduced genome of an enslaved algal nucleus. *Nature* 410(6832), 1091–1096.

Friedman, C. P. (2004). Training the next generation of informaticians: The impact of "BISTI" and bioinformatics – a report from the American College of Medical Informatics. *J. Am. Med. Informatics Assoc.* 11(3), 167–172.

Galagan, J. E., Calvo, S. E., Borkovich, K. A., Selker, E. U., Read, N. D., Jaffe, D., FitzHugh, W., Ma, L. J., Smirnov, S., Purcell, S., Rehman, B., Elkins, T., Engels, R., Wang, S., Nielsen, C. B., Butler, J., Endrizzi, M., Qui, D., Ianakiev, P., Bell-Pedersen, D., Nelson, M. A., Werner-Washburne, M., Seliternnikoff, C. P., Kinsey, J. A., Braun, E. L., Zelter, Z., Schulte, U., Kothe, G. O., Jedd, G., Mewes, W., Staben, C., Marcotte, E., Greenberg, D., Roy, A., Foley, K., Naylor, J., Stange-Thomann, N., Barrett, R., Gnerre, S., Kamal, M., Kamvysselis, M., Mauceli, E., Bielke, C., Rudd,

S., Frishman, D., Krystofova, S., Rasmussen, C., Metzenberg, R. L., Perkins, D. D., Kroken, S., Cogoni, C., Macino, G., Catcheside, D., Li, W., Pratt, R. J., Osmani, S. A., DeSouza, C. P., Glass, L., Orbach, M. J., Berglund, J. A., Voelker, R., Yarden, O., Plamann, M., Seiler, S., Dunlap, J., Radford, A., Aramayo, R., Natvig, D. O., Alex, L. A., Mannhaupt, G., Ebbole, D. J., Freitag, M., Paulsen, I., Sachs, M. S., Lander, E. S., Nusbaum, C., Birren, B. (2003). The genome sequence of the filamentous fungus *Neurospora crassa*. *Nature* 422(6934), 859–868.

Gardner, M. J., Hall, N., Fung, E., White, O., Berriman, M., Hyman, R. W., Carlton, J. M., Pain, A., Nelson, K. E., Bowma, S., Paulsen, I. T., James, K., Eisen, J. A., Rutherford, K., Salzberg, S. L., Craig, A., Kyes, S., Chan, M. S., Nene, V., Shallom, S. J., Suh, B., Peterson, J., Angiuoli, S., Pertea, M., Allen, J., Selengut, J., Haft, D., Mather, M. W., Vaidya, A. B., Martin, D. M., Fairlamb, A. H., Fraunholz, M. J., Roos, D. S., Ralph, S. A., McFadden, G. I., Cummings, L. M., Subramanian, G. M., Mungall, C., Venter, J. C., Carucci, D. J., Hoffman, S. L., Newbold, C., Davis, R. W., Fraser, C. M., Barrell, B. (2002). Genome sequence of the human malaria parasite *Plasmodium falciparum*. *Nature* 419(6906), 498–511.

Goff, S. A., Ricke, D., Lan, T. H., Presting, G., Wang, R., Dunn, M., Glazebrook, J., Sessions, A., Oeller, P., Varma, H., Hadley, D., Hutchison, D., Martin, C., Katagiri, F., Lange, B. M., Moughamer, T., Xia, Y., Budworth, P., Zhong, J., Miguel, T., Paszkowski, U., Zhang, S., Colbert, M., Sun, W. L., Chen, L., Cooper, B., Park, S., Wood, T. C., Mao, L., Quail, P., Wing, R., Dean, R., Yu, Y., Zharkikh, A., Shen, R., Sahasrabudhe, S., Thomas, A., Cannings, R., Gutin, A., Pruss, D., Reid, J., Tavtigian, S., Mitchell, J., Eldredge, G., Scholl, T., Miller, R. M., Bhatnagar, S., Adey, N., Rubano, T., Tusneem, N., Robinson, R., Feldhaus, J., Macalma, T., Oliphant, A., Briggs, S. (2002). A draft sequence of the rice genome (*Oryza sativa* L. ssp. *japonica*). *Science* 296(5565), 92–100.

Goffeau, A. G., Barrell, B. G., Bussey, H., Davis, R. W., Dujon, B., Feldmann, H., Galibert, F., Hoheisel, J. D., Jacq, C., Johnston, M., Louis, E. J., Mewes, H. W., Murakami, Y., Philippsen, P., Tettelin, H., Oliver, S. G. (1996). Life with 6000 genes. *Science* 274(5287), 563–567.

Grodzicker, T., Williams, J., P. Sharp and J. Sambrook, (1974). Physical mapping of temperature-sensitive mutations of adenoviruses. *Cold Spring Harb. Symp. Quant. Biol.* 39, 439–446.

Hane, J. K., Oliver, R. P. (2008). RIPCAL: a tool for alignment-based analysis of repeat-induced point mutations in fungal genomic sequences. *BMC Bioinformatics* 9, 478, doi:10.1186/1471–2105–9–478.

Heller, M. J. (2002). DNA microarray technology: devices, systems, and applications. *Annu. Rev. Biomed. Eng.* 4, 129–153.

Hillis, D. M., Dixon, M. T. (1991). Ribosomal DNA: molecular evolution and phylogenetic inference. *Quarterly Reviews of Biology* 66, 411–453.

Kane, M. D., Jeffrey, L. B. (2007). An information technology emphasis in biomedical informatics education. *J. Biomed. Informatics* 40, 67–72.

Katinka, M. D., Duprat, S., Cornillot, E., Méténier, G., Thomarat, F., Prensier, G., Barbe, V., Peyretaillade, E., Brottier, P., Wincker, P., Delbac, F., Alaoui, H. E., Peyret, P., Saurin, W., Gouy, M., Weissenbach, J., Vivarès, C. P. (2001). Genome sequence and gene compaction of the eukaryote parasite *Encephalitozoon cuniculi*. *Nature* 414(6862), 450–453.

Kim, M., Lee, K. H., Yoon, S. W., Kim, B. S., Chun, J., Yi, H. (2013). Analytical tools and databases for metagenomics in the next-generation sequencing era. *Genomics Inform.* 11(3), 102–113.

Lal, S. B., Pandey, P. K., Rai, P. K., Rai, A., Sharma, A., Chaturvedi, K. K. (2013). Design and development of portal for biological database in agriculture. *Bioinformation* 9(11), 588–598.

Lander, E. S. (2001). Initial sequencing and analysis of the human genome. *Nature* 409(6822), 860–921.

Lander, E. S., Linton, L. M., Birren, B., Nusbaum, C., Zody, M. C., Baldwin, J., Devon, K., Dewar, K., Doyle, M., FitzHugh, W., Funke, R., Gage, D., Harris, K., Heaford, A., Howland, J., Kann, L., Lehoczky, J., LeVine, R., McEwan, P., McKernan, K., Meldrim, J., Mesirov, J. P., Miranda, C., Morris, W., Naylor, J., Raymond, C., Rosetti, M., Santos, R., Sheridan, A., Sougnez, C., Stange-Thomann, N., Stojanovic, N., Subramanian, A., Wyman, D., Rogers, J., Sulston, J., Ainscough, R., Beck, S., Bentley, D., Burton, J., Clee, C., Carter, N., Coulson, A., Deadman, R., Deloukas, P., Dunham, A., Dunham, I., Durbin, R., French, L., Grafham, D., Gregory, S., Hubbard, T., Humphray, S., Hunt, A., Jones, M., Lloyd, C., McMurray, A., Matthews, L., Mercer, S., Milne, S., Mullikin, C. J., Mungall, A., Plumb, R., Ross, M., Shownkeen, R., Sims, S., Waterston, H. R., Wilson, R. K., Hillier, L. W., McPherson, J. D., Marra, M. A., Mardis, E. R., Fulton, L. A., Chinwalla, A. T., Pepin, K. H., Gish, W. R., Chissoe, S. L., Wendl, M. C., Delehaunty, K. D., Miner, T. L., Delehaunty, A., Kramer, J. B., Cook, L. L., Fulton, R. S., Johnson, D. L., Minx, P. J., Clifton, S. W., Hawkins, T., Branscomb, E., Predki, P., Richardson, P., Wenning, S., Slezak, T., Doggett, N., J.-Cheng, F., Olsen, A., Lucas, S., Elkin, C., Uberbacher, E., Frazier, M., Gibbs, R. A., Muzny, D. M., Scherer, S. E., Bouck, J. B., Sodergren, E. J., Worley, K. C., Rives, C. M., Gorrell, J. H., Metzker, M. L., Naylor, S. L., Kucherlapati, R. S., Nelson, D. L., Weinstock, G. M., Sakaki, Y., Fujiyama, A., Hattori, M., Yada, T., Toyoda, A., Itoh, T., Kawagoe, C., Watanabe, H., Totoki, Y., Taylor, T., Weissenbach, J., Heilig, R., Saurin, W., Artiguenave, F., Brottier, P., Bruls, T., Pelletier, E., Robert, C., Wincker, P., Rosenthal, A., Platzer, M., Nyakatura, G., Taudien, S., Rump, A., Smith, D. R., Doucette-Stamm, L., Rubenfield, M., Weinstock, K., Lee, H. M., Dubois, J., Yang, H., Yu, J., Wang, J., Huang, G., Gu, J., Hood, L., Rowen, L., Madan, A., Qin, S., Davis, R. W., Federspiel, N. A., Abola, A. P., Proctor, M. J., Roe, B. A., Chen, F., Pan, H., Ramser, J., Lehrach, H., Reinhardt, R., McCombie, W. R., Bastide, M., Dedhia, N., Blöcker, H., Hornischer, K., Nordsiek, G., Agarwala, R., Aravind, L., Bailey, A. J., Bateman, A., Batzoglou, S., Birney, E., Bork, P., Brown, D. G., Burge, C. B., Cerutti, L., H.-Chen, C., Church, D., Clamp, M., Copley, R. R., Doerks, T., Eddy, S. R., Eichler, E. E., Furey, T. S., Galagan, J., Gilbert, J. G. R., Harmon, C., Hayashizaki, Y., Haussler, D., Hermjakob, H., Hokamp, K., Jang, W., Johnson, L. S., Jones, A. T., Kasif, S., Kaspryzk, A., Kennedy, S., Kent, W. J., Kitts, P., Koonin, E. V., Korf, I., Kulp, D., Lancet, D., Lowe, T. M., McLysaght, A., Mikkelsen, T., Moran, J. V., Mulder, N., Pollara, V. J., Ponting, C. P., Schuler, G., Schultz, J., Slater, G., Smit, A. F. A., Stupka, E., Szustakowki, J., Thierry-Mieg, D., Thierry-Mieg, J., Wagner, L., Wallis, J., Wheeler, R., Williams, A., Wolf, Y. I., Wolfe, H. K., S.-Yang,

P., R.-Yeh, F., Collins, F., Guyer, M. S., Peterson, J., Felsenfeld, A., Wetterstrand, K. A., Myers, R. M., Schmutz, J., Dickson, M., Grimwood, J., Cox, D. R., Olson, M. V., Kaul, R., Raymond, C., Shimizu, N., Kawasaki, K., Minoshima, S., Evans, G. A., Athanasiou, M., Schultz, R., Patrinos, A., Morgan, M. J. (2001). Initial sequencing and analysis of the human genome. *Nature* 409, 860–921.

Louis, E. J. (2011). Population genomics and speciation in yeasts. *Fungal Biology Reviews* 25(3), 136–142.

Martinez, D., Larrondo, L. F., Putnam, N., Gelpke, M. D., Huang, K., Chapman, J., Helfenbein, K. G., Ramaiya, P., Detter, J. C., Larimer, F., Coutinho, P. M., Henrissat, B., Berka, R., Cullen, D., Rokhsar, D. (2004). Genome sequence of the lignocellulose degrading fungus *Phanerochaete chrysosporium* strain RP78. *Nat Biotechnol.* 22(6), 695–700.

McPherson, J. D., Marra, M., Hillier, L., Waterston, R. H., Chinwalla, A., Wallis, J., Sekhon, M., Wylie, K., Mardis, E. R., Wilson, R. K., Fulton, R., Kucaba, T. A., Wagner-McPherson, C., Barbazuk, W. B., Gregory, S. G., Humphray, S. J., French, L., Evans, R. S., Bethel, G., Whittaker, A., Holden, J. L., McCann, O. T., Dunham, A., Soderlund, C., Scott, C. E., Bentley, D. R., Schuler, G., Chen, H. C., Jang, W., Green, E. D., Idol, J. R., Maduro, V. V., Montgomery, K. T., Lee, E., Miller, A., Emerling, S., Kucherlapati, Gibbs, R., Scherer, S., Gorrell, J. H., Sodergren, E., Clerc-Blankenburg, K., Tabor, P., Naylor, S., Garcia, D., Jong, P. J., Catanese, J. J., Nowak, N., Osoegawa, K., Qin, S., Rowen, L., Madan, A., Dors, M., Hood, L., Trask, B., Friedman, C., Massa, H., Cheung, V. G., Kirsch, I. R., Reid, T., Yonescu, R., Weissenbach, J., Bruls, T., Heilig, R., Branscomb, E., Olsen, A., Doggett, N., Cheng, J. F., Hawkins, T., Myers, R. M., Shang, J., Ramirez, L., Schmutz, J., Velasquez, O., Dixon, K., Stone, N. E., Cox, D. R., Haussler, D., Kent, W. J., Furey, T., Rogic, S., Kennedy, S., Jones, S., Rosenthal, A., Wen, G., Schilhabel, M., Gloeckner, G., Nyakatura, G., Siebert, R., Schlegelberger, B., Korenberg, J., Chen, X. N., Fujiyama, A., Hattori, M., Toyoda, A., Yada, T., Park, H. S., Sakaki, Y., Shimizu, N., Asakawa, S., Kawasaki, K., Sasaki, T., Shintani, A., Shimizu, A., Shibuya, K., Kudoh, J., Minoshima, S., Ramser, J., Seranski, P., Hoff, C., Poustka, A., Reinhardt, R., Lehrach, H. (2001). A physical map of the human genome. *Nature* 409(6822), 934–941.

Nasir, A., Kim, K. M., Caetano-Anollés, G. (2014). Global patterns of protein domain gain and loss in superkingdoms. *PLoS Comput. Biol.* 10(1), e1003452, doi:10.1371/journal.pcbi.1003452.

Neufeld, L., Cornog, M. (1999). Database history: from dinosaurs to compact discs. *J. Am. Soc. Inf. Sci.* 37, 183–190.

Nielsen, J. (2001). Metabolic engineering. *Appl Microbiol Biotechnol.* 55, 263–283.

Parekh, S., Vinci, V. A., Strobel, R. J. (2000). Improvement of microbial strains and fermentation processes. *Appl Microbiol Biotechnol.* 54, 287–301.

Rao, V. S., Das, S. K., Rao, V. J., Srinubabu, G. (2008). Recent developments in life sciences research: role of bioinformatics. *African Journal of Biotechnology* 7(5), 495–503.

Redecker, D., Kodner, R., Graham, L. E. (2000). Glomalean fungi from the Ordovician. *Science* 289, 1920–1921.

Santos, C. S. S., Stephanopoulos, G. (2008). Combinatorial engineering of microbes for optimizing cellular phenotype. *Curr Opin Chem Biol.* 12, 168–176.

Stephanopoulos, G. (1999). Metabolic fluxes and metabolic engineering. *Metab Eng.* 1, 1–11.

Thangadurai, D., Sangeetha, J. (2013). Bioinformatics tools for the multilocus phylogenetic analysis of fungi. *In Laboratory protocols in fungal biology: current methods in fungal biology*, ed. by Gupta, V. K., Tuohy, M. G., Ayyachamy, M., Turner, K. M., O'Donovan, A., Springer, New York, pp. 579–592.

Tibayrenc, M. (2005). Bridging the gap between molecular epidemiologists and evolutionists. *Trends Microbiol.* 13, 575–580.

Venter, J. (2001). The sequence of the human genome. *Science* 291(5507), 1304–1351.

Verdoes, J. C., Punt, P. J., van den Hondel, C. A. M. J. J. (1995). Molecular genetic strain improvement for the overproduction of fungal proteins by filamentous fungi. *Applied Microbiology and Biotechnology* 43(2), 195–205.

Wafaa, H. M., Mohamed, H. A. A. (2007). Biotechnological aspects of microorganisms used in plant biological control. *World J. Agric. Sci.* 3, 771–776.

Wayne, M., Wishart, D. S. (2007). Computational systems biology in drug discovery and development: methods and applications. *Drug. Discov. Rev.* 12, 295–303.

Wood, V., Gwilliam, R., Rajandream, M. A., Lyne, M., Lyne, R., Stewart, A., Sgouros, J., Peat, N., Hayles, J., Baker, S., Basham, D., Bowman, S., Brooks, K., Brown, D., Brown, S., Chillingworth, T., Churcher, C., Collins, M., Connor, R., Cronin, A., Davis, P., Feltwell, T., Fraser, A., Gentles, S., Goble, A., Hamlin, N., Harris, D., Hidalgo, J., Hodgson, G., Holroyd, S., Hornsby, T., Howarth, S., Huckle, E. J., Hunt, S., Jagels, K., James, K., Jones, L., Jones, M., Leather, S., McDonald, S., McLean, J., Mooney, P., Moule, S., Mungall, K., Murphy, L., Niblett, D., Dell, C., Oliver, K., O'Neil, S., Pearson, D., Quail, M. A., Rabbinowitsch, E., Rutherford, K., Rutter, S., Saunders, D., Seeger, K., Sharp, S., Skelton, J., Simmonds, M., Squares, R., Squares, S., Stevens, K., Taylor, K., Taylor, R. G., Tivey, A., Walsh, S., Warren, T., Whitehead, S., Woodward, J., Volckaert, G., Aert, R., Robben, J., Grymonprez, B., Weltjens, I., Vanstreels, E., Rieger, M., Schäfer, M., Müller-Auer, S., Gabel, C., Fuchs, M., Düsterhöft, A., Fritzc, C., Holzer, E., Moestl, D., Hilbert, H., Borzym, K., Langer, I., Beck, A., Lehrach, H., Reinhardt, R., Pohl, T. M., Eger, P., Zimmermann, W., Wedler, H., Wambutt, R., Purnelle, B., Goffeau, A., Cadieu, E., Dréano, S., Gloux, S., Lelaure, V., Mottier, S., Galibert, F., Aves, S. J., Xiang, Z., Hunt, C., Moore, K., Hurst, S. M., Lucas, M., Rochet, M., Gaillardin, C., Tallada, V. A., Garzon, A., Thode, G., Daga, R. R., Cruzado, L., Jimenez, J., Sánchez, M., Rey, F. D., Benito, J., Domínguez, A., Revuelta, J. L., Moreno, S., Armstrong, J., Forsburg, S. L., Cerutti, L., Lowe, T., McCombie, W. R., Paulsen, I., Potashkin, J., Shpakovski, G. V., Ussery, D., Barrell, B. G., Nurse, P. (2002). The genome sequence of *Schizosaccharomyces pombe. Nature* 415(6874), 871–880.

Wooley, J. C., Lin, H. S. (2005). *Catalyzing inquiry at the interface of computing and biology.* National Academies Press, Washington, USA.

Zauhar, J. R. (2001). University bioinformatics programs on the rise. *Nature Biotechnology* 19, 285–286.

CHAPTER 2

FUNGAL IDENTIFICATION: CONVENTIONAL APPROACHES AND CURRENT SCENARIO

CONTENTS

2.1   INTRODUCTION

Previously, taxonomic identification of fungi was based on the observations of morphological traits, such as cultural morphologies, including colony and color characteristics on specific culture media, its size and shape, development of sexual, asexual spores, spore-forming structures and physiological characteristics such as the ability to utilize various compounds as source of nitrogen and carbon (Felsenstein, 1985). These methods tend to be time consuming, laborious and may take several days for isolation. Moreover, fungal species differ in their carbon utilization and cannot be cultured on a given medium, which leads inaccurate analysis of the species that may not belong to the true fungal community (Wu et al., 2003). The uncultured and non-viable spores can possibly be allergenic and cause health issues. For instance, the conidia of

Stachybotrys chartarum rapidly lose viability when it comes in contact with air, without losing its toxigenicity (Hibbett et al., 2007). Thus, rapid detection of fungi in a given environment is essential for monitoring the exposure risk and for developing precautionary measures for public health safety (Zeng et al., 2003). Biochemical markers has been assumed reliable and is tried out extensively, but no encouraging results are obtained, as they are often sex limited, age-dependent and are significantly influenced by the environment (Alim et al., 2011). Sometimes, the various genotypic classes are indistinguishable at the phenotypic level owing to dominance effect. Moreover, these markers show variability in their coding sequences that constitute less than 10 per cent of the total genome (Zwickl, 2006).

Fungal cells contain DNA as chromosomes in the nucleus, and in the mitochondria, a distinguishing characteristic of eukaryotes. Chromosomes contain innumerable sequences, of which some are organized into genes, others as merely flanking or spacer regions (Tautz, 1989). An eukaryotic cell consists of one or more nucleus containing single and multiple copy sequences and many mitochondria, each containing sequences identical to each other.

2.2   DNA-DNA HYBRIDIZATION

A method first employed for the identification of fungi is DNA hybridization. During this process, the total DNA from two fungal species was extracted. Then the DNA was denatured to obtain single strands and mixed together. The proportion of DNA from one organism that was able to form a double stranded molecule with the DNA from the other organism was then determined. This gave an overall similarity of the DNA sequences between the two organisms. In general, this practice was relatively complex, as the entire DNA sequences had to be labeled, usually with radioactive labels to allow identification of the mixed double strands. DNA-DNA hybridization has two basic methods: thermal stability and cross hybridization. But for fungi cross hybridization between the total DNA extracts of the organism is preferred to thermal stability testing method. The binding specificity between the two strands was also dependent on temperature and salt concentration, and so reaction conditions and reagents were crucial.

If DNA/DNA values of greater than 70%, then closely related strains; values of below 40% showed that the strains less related (Bruns et al., 1991; Royse et al., 1993). This method has restricted its use only within the yeast genera. The major setback with fungal species is that due to its genome size considerable background similarity was obtained which resulted in much higher values required to define species. Due to few clear cut-off values, it was difficult to differentiate between closely related species and less related species.

2.3 ALLOZYMES

The first molecular markers to be developed were allozymes (Schlötterer, 2004) and they are the enzymes found in biological system involved in DNA repair and replication. They are the variant forms of an enzyme that are coded by different alleles present on the same loci. These enzymes are structurally different due to difference in amino acid sequence between phyla. Allozymes are widely used in fungal identification as molecular markers (Urbanelli et al., 2003); they compare between different species by selecting one that is as variable as possible nevertheless they are present in all the organisms and be comparable to amino acid sequence of the enzyme in the species, more amino acid similarities should be seen between evolutionarily related species and fewer between those that are less well related. The less, well conserved the enzyme is, the more amino acid differences will be present in even closely related species. Because of amino acid charge differences, allozymes can be differentiated by their relative migration speed during gel electrophoresis. Many enzymes are invariant within populations (or even between species and higher taxa), and most polymorphic enzymes have only a few variants (generally two) (Mueller, 1999). Although this limits the power of allozyme analysis to resolve genetic differences, allozymes are time and cost efficient for research, requiring only a few polymorphic markers, constraints in using allozymes are moderate number of markers, inadequate for investigating considerable portions of the genome and complications in sharing the experimental procedure between research laboratories.

2.4   POLYMERASE CHAIN REACTION

Polymerase chain reaction (PCR) was first described in the mid 1980s (Saiki et al., 1985), Kary Mullis first demonstrated PCR and is a process that allows many copies to be made of a DNA region. The PCR process can be performed on tiny samples with DNA concentrations of nanograms or less. The PCR process works by making the target DNA single stranded, and then attaching small synthetic primers to each end of the region of interest. The region of DNA between the primers is then constructed from the target DNA with individual phosphorylated nucleotides by a specific enzyme DNA polymerase. The primary condition for the reaction process is temperature, as a high temperature >90°C is needed in order to denature DNA molecule into single strands, and a lower temperature 35–55°C is required to allow the primers to bind. A further step is required for the synthesis of the DNA, and this takes place at the optimum temperature for the DNA polymerase. Each step in the process typically takes between 30 and 120 seconds, and the complete cycle results in a duplication of the original DNA sequence. Further, doubling samples rapidly increase the amount of the DNA region of interest (typically 25–45 cycles). The DNA polymerase enzyme used therefore needs to be thermostable in order to withstand repeated cycles of heating at more than 90°C, and it was the discovery of heat resistant enzymes in thermophilic bacteria such as *Thermus aquaticus*, that allowed the process to become a viable practical method (Saiki et al., 1988).

There have been many different applications of PCR methods in mycology and extensive information has been discussed by Edel (1998). It is evident from above statements that, the PCR process is a way of obtaining large quantities of a particular region of DNA. It is not however an analysis tool, and so after PCR it is necessary to determine features of the amplified DNA. In the simplest form, the presence/absence of an amplified product may be a final result, such as in diagnostics where the primers used will only bind with the DNA of certain organisms. In other cases the size and number of products may be the end result, such as in cases where much generalized primers are used to generate a DNA fingerprint (Mullis et al., 1986). More commonly however some degree of sequence information is usually derived from the PCR amplified product, either through restriction enzyme analysis or sequencing. Automated DNA sequencing is now available through systems based either on electrophoretic separation on gels or on microcapillary electrophoresis. PCR methods are used to generate dye-labeled fragments that are detected by either variable or fixed wavelength lasers. The time taken to obtain a sequence, the number of samples that can be analyzed, and the maximum length of sequence will depend on a number of factors, including the type of sequencer, but in many applications automated sequencers can provide around 1000bp of sequence from overnight operations (Mullis and Faloona, 1987). PCR-based genetic marker for detection and infection frequency analysis of the biocontrol fungus *Chondrostereum purpureum* on Sitka Alder and Trembling Aspen was obtained. Food and forensic laboratories are also using PCR for identification of pathogenic fungi (Fujita and Silver, 1994).

KEYWORDS

- Allozymes
- Asexual spores
- Automated DNA sequencing
- Carbon utilization
- Cultural morphologies
- DNA fingerprint
- DNA polymerase
- DNA repair
- DNA-DNA hybridization
- Dye-labeled fragments
- Fungal community
- Gel electrophoresis
- Heat resistant enzymes
- Microcapillary electrophoresis
- Morphological traits

- **PCR-based genetic marker**
- **Polymerase chain reaction**
- **Polymorphic enzymes**
- **Polymorphic markers**
- **Restriction enzyme analysis**
- **Spore-forming structures**
- **Thermophilic bacteria**
- *Thermus aquaticus*

REFERENCES

Alim, M. A., Sun, D. X., Zhang, Y., Faruque, M. O. (2011). Genetic markers and their application in buffalo production. *Journal of Animal and Veterinary Advances* 10(14), 1789–1800.

Bruns, T. D., White, T. J., Taylor, J. W. (1991). Fungal molecular systematics. *Annu. Rev. Ecol. Syst.* 22, 525–564.

Edel, V. (1998). PCR in mycology: an overview. *In Application of PCR in Mycology*, ed. by Bridge, P. D., Arora, D. K., Reddy, C. A., Elander, R. P., CAB International, Oxford, UK, pp. 1–20.

Felsenstein, J. (1985). Confidence limits on phylogenies: An approach using the bootstrap. *Evolution* 39(4), 783–791.

Hibbett, D. S., Binder, M., Bischoff, J. F., Blackwell, M., Cannon, P. F., Eriksson, O. E., Huhndorf, S., James, T., Kirk, P. M., Lücking, R., Lumbsch, H. T., Lutzoni, F., Matheny, P. B., Mclaughlin, D. J., Powell, M. J., Redhead, S., Schoch, C. L., Spatafora, J. W., Stalpers, J. A., Vilgalys, R., Aime, M. C., Aptroot, A., Bauer, R., Begerow, D., Benny, G. L., Astlebury, L. A., Crous, P. W., Dai, Y. C., Gams, W., Geiser, D. M., Griffith, G. W., Gueidan, C., Hawksworth, D. L., Hestmark, G., Hosaka, K., Humber, R. A., Hyde, K. D., Ironside, J. E., Kõljalg, U., Kurtzman, K. C. P., Larsson, H., Lichtwardt, R., Longcore, J., Miądlikowska, J., Miller, A., Moncalvo, J. M., Mozley-Standridge, S., Oberwinkler, F., Parmasto, E., Reeb, V., Rogers, J. D., Roux, C., Ryvarden, L., Sampaio, J. P., Schüßler, A., Sugiyama, J., Thorn, R. G., Tibell, L., Untereiner, W. A., Walker, C., Wang, Z., Weir, A., Weiss, M., White, M. M., Winka, K., Yao, Y. J., Zhang, N. (2007). A higher-level phylogenetic classification of the fungi. *Mycological Research* 111, 509–547.

Mueller, U. G., Wolfenbarger, L. L. (1999). AFLP genotyping and fingerprinting. *Tree* 14, 389–394.

Mullis, K., Faloona, E., Scharf, S., Saiki, R., Horn, G., Erlich, H. (1986). Specific enzymatic amplification of DNA *in vitro*: the polymerase chain reaction. *Cold Spring Harbor Symp. Quant. Biol.* 51, 263–273.

Mullis, K. B., Faloona, F. A. (1987). Specific synthesis of DNA *in vitro* via a polymerase-catalyzed chain reaction. *Methods Enzymol.* 155, 335–350.

Royse, D. J., Bunyard, B. A., Nicholson, M. S. (1993). Molecular genetic analysis of diversity in populations of edible mushrooms. *In Mushroom Biology and Mushroom Products*, ed. by Chang, S. T., Buswell, J. A., Chiu, S., The Chinese University Press, Hong Kong, pp. 49–54.

Saiki, R., Gelfand, D., Stoffel, S., Scharf, S., Higuchi, R., Horn, G., Mullis, K., Erlich, H. (1988). Primer-directed enzymatic amplification of DNA with a thermostable DNA polymerase. *Science* 239, 487–491.

Saiki, R. K., Scharf, S., Faloona, F., Mullis, K. B., Horn, G. T., Erlich, H. A., Arnheim, N. (1985). Enzymatic amplification of beta-globin genomic sequences and restriction site analysis for diagnosis of sickle cell anemia. *Science* 230(4732), 1350–1354.

Schlötterer, C. (2004). The evolution of molecular markers-just a matter of fashion? *Nat. Rev. Genet.* 5, 63–69.

Tautz, D. (1989). Hyper variability of simple sequences as a general source for polymorphic DNA markers. *Nucleic Acids Research* 17, 6463–6471.

Urbanelli, S., Rosa, V. D., Fanelli, C., Fabbri, A. A., Reverberi, M. (2003). Genetic diversity and population structure of the Italian fungi belonging to the taxa *Pleurotus eryngii* (DC.:Fr.) Quèl and *P. ferulae* (DC.:Fr.) Quèl. *Heredity* 90, 253–259.

Wu, Z., Tsumura, Y., Blomquist, G., Wang, X.-R. (2003). 18S rRNA gene variation among common airborne fungi, and development of specific oligonucleotide probes for the detection of fungal isolates. *Appl. Environ. Microbiol.* 69(9), 5389–5397.

Zeng, Q.-Y., Wang, X.-R., Blomquist, G. (2003). Development of mitochondrial SSU rDNA-based oligonucleotide probes for specific detection of common airborne fungi. *Molecular and Cellular Probes* 17, 281–288.

Zwickl, D. J. (2006). Genetic algorithm approaches for the phylogenetic analysis of large biological sequence datasets under the maximum likelihood criterion. *PhD Dissertation*, The University of Texas at Austin, USA.

CHAPTER 3

MOLECULAR TOOLS FOR IDENTIFICATION OF FUNGI

CONTENTS

3.1   INTRODUCTION

The genome of all the known species on earth is almost constituted by repetitive DNA. Genetic identification of all these species is possible only after the discovery of these highly polymorphic of alleles, which constitute 30–90% of the genomes. Since several hundred alleles are combined to form genetic loci in DNA regions, they are not similar across species and varies with respect to length, sequence and tandem arrays of distribution. Normally, the inherited mutations that occur in the repetitive DNA region are more responsible for diversity and evolution. In recent years, diverse groups of hybridization and PCR based molecular, genomic and proteomic markers available are heavily been used for identification of fungi (Ellegren, 2004; Table 3.1).

TABLE 3.1   Different Molecular Markers Used in the Systematics and Taxonomy of Fungi

Fungi	Molecular markers	References
A. fumigates	Multiple-locus variable number tandem repeats	Thierry et al. (2010)
Aspergillus sojae	RAPD	Leach et al. (1986)
Basidiobolus ranarum	PCR-amplified ribosomal RNA genes	Chakrabarti et al. (2003)
Bipolaris oryzae	Ribosomal region, protein coding genes	Bass et al. (1992)
Chionasphaera apabasidialis	Tubulin	Berbee and Taylor (1992)
Colletotrichum acutatum	mtDNA, RFLP	Adaskeveg and Hartin (1997)
Fusarium solani	RFLP, RAPD, ISSR, ITS	Bakshi (1955)
Ganoderma lucidum	RAPD	Nogami (1987)
Glomus intraradices	mt-large subunit	Almeida and Schenk (1990)
Glomus proliferum	mt-large subunit	Almeida and Schenk (1990)
Heterobasidion annosum	Protein coding genes	Adams et al. (1991)
Naohidea sebacea	Tubulin	Berbee and Taylor (1992)
Nomuraea rileyi	RAPD, AFLP, ITS	Grooters and Melaney (2002)
Paxillus involutus	ITS	Wright et al. (2001)
Pisolithus tinctorius	RAPD	Agerer (1991)
Scaulochytrium aggregatum	PCR-amplified ribosomal RNA genes	Chakrabarti et al. (2003)

3.2   RESTRICTION FRAGMENT LENGTH POLYMORPHISM

Restriction fragment length polymorphism (RFLP) is one of the best techniques that uses specific enzymes called restriction endonucleases (REs). They are DNA-digesting enzymes that recognize particular sequences within the DNA and cut the DNA at specific restriction sites. Hence, they are commonly called as molecular scissors due to their functionality. The target sequences are palindromic, wherein the sequences are read same on the complementary strands in the opposite direction, and so, the DNA is cut at the same location on both the strands. There are different types of restriction endonucleases which vary from four bases, upto recognition sites of twelve or more bases. Therefore, if DNA is treated with REs that recognizes a sequence that occurs rarely, then the DNA will be cut into a few large fragments, whereas if the enzyme recognizes a sequence that occurs many times, then the DNA will be cut into many small fragments, and these fragments are blotted onto a nitrocellulose membrane (Cooley, 1992). Specific banding patterns are then visualized by hybridization with labeled probe. This technique is called restriction fragment length polymorphism analysis. This technique can give rapid information about the location of restriction enzyme sites and the size of DNA of interest. Restriction digestion of genomic DNA yields large number of fragments and RFLP analysis normally results in these fragments being visualized as an extended smear in the gel. Moreover, fragments of DNA that appears as distinct bands have been used widely to differentiate fungal isolates of the same species (Coddington et al., 1987). Probes are short segments of DNA and are of specific genes. These DNA probes can be obtained by various methods in which, cloning is a common method employed. In cloning, once restriction digestion of the DNA strand by restriction endonucleases is completed, the short DNA fragments are inserted into bacterial plasmids, bacteriophages or artificial vectors. Thus, recombinant molecules obtained are multiplied through rapid bacterial growth; the probes are then labeled with radioactive labels or with a particular enzyme. These single locus species specific probes (0.5–3.0 kb) are obtained from cDNA and genomic libraries. Genomic libraries are easy to construct and virtually all types of sequences are included, on the contrary, cDNA libraries are popular since actual genes are analyzed and they hardly carry any repetitive sequence (Mavridou and Typas, 1998).

The selection of desirable RFLP probe source depends upon the kind of applications. In general, genomic library probes exhibit greater variability than cDNA library probes. In contrast, cDNA library probes may also show greater variability due to their ability to detect regions with flanking genes and introns, apart from detecting variation in coding regions of the corresponding genes. RFLP markers, being co-dominant are reliable in linkage analysis and used in the construction of genetic maps. Also, they can recognize coupling phase of DNA sequences, as DNA fragments from all homologous chromosomes are distinguished. However, a drawback of this method is that their utility has been hampered due to the large amount of DNA required for restriction digestion and southern blotting. Furthermore, a need of radioactive isotope makes the analysis comparatively expensive and unsafe. The assay is tedious and elaborate and only one out of several markers may be polymorphic, which is highly unsuitable principally for crosses between closely related species. Moreover, RFLP marker shows their inability to recognize point mutations prevailing within the regions where they detect polymorphism (Toda et al., 1998).

3.3   RANDOM AMPLIFIED POLYMORPHIC DNA

The PCR-based random amplified polymorphic DNA (RAPD) analysis was developed by Welsh and McClelland (1991). This technique works on the principle of DNA amplification, and these are very quick and easy to develop due to the arbitrary sequence of the primers. It gives accurate results even with small amount of DNA samples and works without a need of radioisotope. The development of RAPD markers is easy and less time consuming, as it does not require any specific knowledge of the DNA sequence of the target organism. It involves the detection of nucleotide sequence polymorphisms in DNA by using a single primer of arbitrary nucleotide sequence. Between 100 and 4000 base pair standard amplicons of variable length pairs are created, each primer controls amplification of several distinct loci in the genome, molding the assay useful for systematic screening of nucleotide sequence polymorphism between individuals. Yet, due to the speculative nature of DNA amplification with random sequence primers, it is essential to optimize and retain stable reaction conditions and annealing temperature for reproducible DNA amplification (Sharon et al., 2006). RFLPs are codominant markers, accordingly have limitations in their use as markers for mapping genes, which can be overcome to some level by choosing those markers that are combined in coupling. RAPD assay is being used by number of scientific groups as a systematic tool for identification of marker linked to agronomically important genotypes. Applying this technique for identification and differentiation of yeast belonging to the genera *Saccharomyces* and *Zygosaccharomyces*, five primers were used and results obtained proved that RAPD analysis is a powerful technique (Paffetti et al., 1995). RAPD technique is being used to generate molecular markers which are useful in identifying fungi, and differentiate them at the species, subspecies or strain level, and to study the genetic polymorphism of genotypes at molecular level. Geisen (1995) used RAPD analysis to characterize sixteen strains of *P. nalgiovense* at genetic level and found that *P. nalgiovense* is closely related to *P. chrysogenum*. Characterization of *Aspergillus chevalieri*, *Aspergillus nidulans*, *Aspergillus tetrazonus* and their teleomorphs was successfully done with RAPD analysis using two primers. RAPD used to identify races of 1, 2, 4, and 8 of *Fusarium oxysporum* f. spp. *dianthi* in Italy (Migheli et al., 1998). Cobb and Clarkson (1993) studied the DNA polymorphism among insect pathogenic fungus *Metarhizium anisopliae* and *M. flavoviride* and Caldeira et al. (2009) characterized *Amanita ponderosa* (mushroom). RAPD is successfully being used in gene mapping, genetic identity, taxonomic identification, genotoxicity and carcinogenesis studies.

Some modifications in RAPD resulted in new techniques like DNA amplification fingerprinting (DAF) and arbitrary primed polymerase chain reaction (AP-PCR). In DAF, single arbitrary primers as short as five bases are used to amplify DNA sequences using PCR. A spectrum of products is obtained containing simple and complex patterns; simple patterns are used as markers for genetic mapping, whereas more complex patterns are used for DNA fingerprinting. The resulting band patterns are reproducible and band pattern analysis is carried out using polyacrylamide gel electrophoresis and silver staining. The parameters are required to be optimized precisely; however, DAF is an automated method and the primers are fluorescently tagged for easy and rapid identification of amplified DNA products (Tinker et al., 1993). DAF profiles can be tuned by employing modifications like predigesting of DNA template. The application of DAF is found in genetic mapping and genetic typing of fungi. Arbitrary primed polymerase reaction (AP-PCR) is a unique type of RAPD which includes single primers of length 10–50 bases and generates discrete amplification patterns of genomic DNA in PCR. Annealing is carried out in nonstringent conditions for first two cycle of PCR and the PCR products obtained are structurally similar to RAPD. AP-PCR is not as popular as DAF as it involves autoradiography. However, recently it has been simplified by including agarose gel electrophoresis for separating fragments and ethidium bromide staining for visualization.

3.4 AMPLIFIED FRAGMENT LENGTH POLYMORPHISM

Amplified fragment length polymorphism (AFLP) technique is a combination of RFLP and PCR originally developed by Zabeau and Vos (1993). This technique is useful in understanding polymorphism between closely related genotypes based on the detection of genomic restriction fragments by PCR amplification. AFLP has high multiplex ratio and large number of polymorphism making it unique among several other molecular techniques. The AFLP fingerprints can be produced without any prior genetic sequence knowledge using a limited set of primers and adapters. The number of fragments detected in a single reaction can be specifically selected by primer sets (Strommer et al., 2002). AFLP technique is reliable, after all rigorous reaction conditions are used for primer annealing. In this technique, DNA is digested with a rare cutting and commonly cutting restriction enzymes simultaneously, such as MseI and EcoRI, and ligated with specific oligonucleotide adaptors to restriction fragments, then selectively amplifying these fragments with specifically designed primers. For amplification purpose the 5′ end of the primer is designed such that it would contain both the restriction enzyme sites on either sides of the fragment complementary to respective adapters, while the 3′ region extend for a few randomly selected nucleotides into the restriction fragments. After amplification, separation of fragments will be carried out on agarose gel electrophoresis followed by visualization using autoradiography, silver staining or fluorescent dyes. AFLPs are exclusively used as tools for DNA fingerprinting and also for cloning, genetic mapping and linkage analysis of variety-specific genomic DNA sequences. Similar to RAPDs, the bands of interest obtained by AFLP can be converted into sequence-characterized amplified regions (SCAR). Thus, AFLP provides a newly developed, most vital tool for variety of applications (Voort et al., 1997). AFLP analysis executes unique fingerprints irrespective of origin and genome complexity. Limitation of using AFLP is the difficulty in developing locus-specific marker from individual fragments.

3.5 SEQUENCE-RELATED AMPLIFIED POLYMORPHISM

SRAP, a modified marker technology termed as sequence-related amplified polymorphism is similar to RAPD. In studying the genetic diversity of fungi compared to other molecular techniques, SRAP markers has the advantage of producing high resolution, good reproducibility and more genetic information. SRAP had been applied abundantly in gene mapping, genetic diversity analysis and comparative genetics of different species (Li and Quiros, 2001). Furthermore, in genetic diversity analysis, the information derived from SRAP marker was more concordant to the morphological variability and to the evolutionary history of the morphotypes than any other molecular markers (Ferriol et al., 2003). SRAP technique is a PCR-based marker system employing a combination of two primers, a forward primer and reverse primer of 17 bases and 18 bases respectively, which consisted of preferential amplification of open reading frames (ORFs). The forward primers are fixed sequence containing 14 bases of G and C rich region at 5' end and three specific bases at 3' end. The first 10 bases at the 5' end are called filler bases, which are of no specific function, followed by CCGG bases and then three specific bases at 3' end. In reverse primer, the filler sequence is followed by AATT bases at the 5' end and three selective bases at 3' end. The principle laid for the construction of forward and reverse primers is to prevent the formation of hairpins or other secondary structures, 40–50% of GC content and the filler sequences in the forward and reverse primers should not be similar to each other. The forward primers prefer amplification of exonic regions and reverse primers preferably amplify intronic regions. The polymorphism observed originated fundamentally due to the variation in the length of introns, promoters and spacers, both among individuals and among species (Lin et al., 2005).

SRAP technique is used for the genetic identification of fungi. This technique is simple and efficient and offers several advantages, such as rational throughput rate, identification of numerous codominant markers, and easy isolation of bands for sequencing. It mainly targets ORFs, which can be used for various applications in identification of fungi including SCAR marker, map construction and gene chip. SRAP markers can provide more precise and accurate data on population genetic diversity over traditional methods. Sun et al. (2006) performed SRAP analysis for identification of genetic diversity in endophytic fungi from *Taxus*, and suggested that this powerful technique could be used for the study of endophytic fungi. SRAPs have also been demonstrated to be very powerful tool in many fungal species including *Pleurotus citrinopileatus, Pleurotus geesteranus, Puccinia striiformis, Monascus* sp. (Shao et al., 2011) and *Fusarium oxysporum* (Mutlu et al., 2008).

3.6 ALLELE-SPECIFIC ASSOCIATED PRIMERS

An allele-specific associated primer (ASAP) method is another type of PCR technique which is fast, high multiplex assay that commonly shows up significant polymorphisms. This technique does not require prior knowledge about the genome makeup of the organism (Vos et al., 1995). To obtain an allele-specific marker, specific allele of interest, either in homozygous or heterozygous conditions in an individual is sequenced and distinct primers are designed for the amplification of DNA template to generate single fragment at severe annealing temperatures. These markers are identified by the sequence of the decamer oligo derived from normal RAPD. Allele specific associated primers are used to generate a single DNA fragment from a DNA template extracted under alkaline conditions.

3.7 SINGLE STRAND CONFORMATION POLYMORPHISM

The single-strand conformation polymorphism (SSCP) technique was discovered by Orita et al. (1989). The SSCP technique has potential to identify most sequence variations in a single strand of DNA, consistently between 150 and 250 nucleotides in length. This is substantial and required for gene probing particularly for recognition of point mutations and typing of

DNA polymorphism. SSCP diagnosis heterozygosity of DNA segments of identical molecular weight and can even detect modified nucleotide bases, as the mobility of the single-stranded DNA alters with change in its GC proportions due to its structural change. Kong et al. (2003) characterized 29 species and 282 isolates of *Phytophthora* sp. through SSCP of ribosomal DNA, and arrived at the conclusion that SSCP could be useful for taxonomic, genetic and ecological studies. The SSCP technique facilitates the revelation of both familiar and unnoted single point mutations and polymorphisms in PCR products. Optimization of SSCP analysis to detect the maximum number of mutations requires electrophoresis under carefully controlled conditions at different temperatures and using different gels.

3.8   EXPRESSED SEQUENCE TAGS

Expressed sequence tags (ESTs) are short DNA sequences, consistent with a complimentary DNA (cDNA) fragment, which may be expressed in a cell at a specific given time. ESTs are DNA fragments (300–500 bp) that are reverse-transcribed from cellular mRNA molecules (Brenner et al., 2003). They are developed by extensive single-pass sequencing of adventitiously selected cDNA clones and have proven to be orderly and rapid means of ascertaining novel genes. ESTs are currently considered to be rapid and efficient method of profiling genes expressed in heterogeneous tissues, cell types or developmental stages. One of the fascinating appliances of ESTs database (dbEST) is gene discovery, where, a lot of new genes are found with a protein or DNA sequence by querying the dbEST. Moreover, ESTs are popularly used in full genome sequencing, for identifying active genes and mapping programs are underway for number of organisms. Thus, helping in identification of diagnostic markers, for example, Stukenbrock et al. (2005) isolated and characterized EST-derived microsatellite loci from the fungal wheat pathogen *Phaeosphaeria nodorum*.

3.9   MICROSATELLITES AND MINISATELLITES

RFLP probes especially associated to a desired trait can be altered into PCR-based STS markers derived from nucleotide sequence of the probe giving polymorphic band pattern for obtaining specific amplicon by using this technique; tedious hybridization procedures involved in RFLP assay can be overcome. This approach is extremely useful in studying the relationship between different species. When these markers are linked to some specific traits, they can be easily integrated into plant breeding programs for marker-assisted selection of the trait of interest. For example, Brisbane et al. (1995) identified *Rhizoctonia solani* AG 4 and *Rhizoctonia solani* AG 8 infecting wheat in South Australia by STS markers. STS have several advantages, in that they are co-dominant, tend to be more reproducible, and they can distinguish between homozygotes and heterozygotes.

The word microsatellite was coined by Litt and Lutty, while the term minisatellites was introduced by Jeffrey, both are multi-locus probes developing complex banding patterns and are usually non-species specific occurring universally. They essentially belong to the repetitive DNA family. Fingerprints generated by these probes are also known as oligonucleotide fingerprints. This technique has been derived from RFLP and identified fragments are visualized by hybridization with a labeled micro- or minisatellite probe. Minisatellites are tandem repeats including a monomer repeat length of about eleven to sixty base pairs, while microsatellites or short tandem repeats/simple sequence repeats (STRs/SSRs) of one to six base pair long. These loci contain tandem repeats that vary in the number of repeat units between genotypes and are referred to as variable number tandem repeats (VNTRs), e.g., a single locus that contains variable number of tandem repeats between individuals or hypervariable regions (HVRs). Microsatellites and minisatellites are an excellent molecular marker system producing complex banding patterns by simultaneous detection of multiple DNA loci. Some of the prominent features of these markers are that they are dominant fingerprinting markers and co-dominant sequence tagged microsatellites markers (STMS).

3.10   NUCLEAR rDNA SEQUENCES

Nuclear internal transcribed spacer (ITS) regions are one of the most indispensable tools for phylogenetic inference at genus, species and even within species to identify geographic races (Kuninaga et al., 1997). ITS, refers to the piece of structural ribosomal RNA consisting of 5'-ETS, 18S rRNA, ITS1, 5.8S rRNA, ITS2, 26S rRNA and the 3'-ETS on a common precursor transcript. ETS and ITS pieces are excised and rapidly degraded as non-functional byproducts during rRNA maturation. Genes encoding ribosomal RNA and spacers occur in tandem repeats, each separated by regions of non-transcribed DNA termed as intergenic spacer (IGS) or non-transcribed spacer (NTS). Sequence comparison of the ITS region is widely used in taxonomy and molecular phylogeny (Gardes et al., 1993). For example, Wang et al. (2006) studied the evolution of helotialean fungi by using sequences of rDNA regions namely SSU, LSU, and 5.8S rDNA, and constructed nuclear rDNA phylogeny. Giachini et al. (2010) studied the phylogenetic relationship among Gomphales based on nuc-25S-rDNA, mit-12S-rDNA, and mit-atp6-DNA combined sequences. Choi et al. (2006) studied the genetic diversity among *Albugo candida* complex using ITS rDNA and COX2 mtDNA sequences. Tanabe et al. (2000) analyzed 18rDNA sequences of Zygomycota and a molecular phylogeny was deduced. Rouland-Lefevre et al. (2002) analyzed Termitomyces species by the sequencing of their internal transcriber spacer region (ITS1–5.8S-ITS2).

3.11 MITOCHONDRIAL DNA

In fungi, mitochondrial DNA differs from nuclear DNA in its location, genome size and G+C content (Grossman et al., 1985). Mitochondrial DNA can therefore be separated from nuclear DNA either by extraction of complete mitochondria, followed by DNA extraction, or by total DNA extraction and fractionation, usually by density gradient centrifugation. Due to its size, digestion of mitochondrial DNA with restriction enzymes can lead to a relatively small number of fragments, and so direct RFLPs can be generated (Marriott et al., 1984). Mitochondrial DNA in fungi varies widely in size, and generally ranges between 30 thousand and 120 thousand bases in length. In fungi, the regions of mitochondrial DNA genes are well conserved, but the mitochondrial DNA can contain large amounts of non-coding DNA sequences referred to as introns (Charter et al., 1996). Much of the variation in mitochondrial DNA is found in these regions, and mitochondrial analysis has been used quite extensively in mycology. Although a relatively old method, this can still be a useful way of characterizing closely related isolates or species (Varga et al., 1993). It must however be remembered that the rate of evolution of mitochondrial DNA may be quite different from that of nuclear DNA, and so conclusions derived from mitochondrial sequences may not correspond with those from nuclear DNA. Zeng et al. (2003) developed mitochondrial SSU rDNA-based oligonucleotide probes for specific detection of common airborne fungi.

3.12 RIBOSOMAL RNA GENE CLUSTER

The rRNA cluster has been used extensively in evolutionary and systematic studies, and also for the development of molecular diagnostic tools in medicine, agriculture and ecology (Bruns et al., 1991). The cluster consists of three major genes that code for large, small and 5.8S ribosomal subunits which are present between intergenic spacer (IGS), non-transcribed spacer (NTS), externally transcribed spacer (ETS) and internally transcribed spacer (ITS). The whole gene cluster is repetitive along the chromosome, wherein individual clusters are being separated by intergenic spacer sequences (Bowman et al., 1992). Among the ribosomal RNA gene clusters, large subunit carries a wide range of informative characters for phylogenetic studies at the higher taxonomic levels. Since the large subunit and the small subunit genes contain both conserved and variable regions, probes and primers derived from these conserved regions can be used for many phylogenetic studies among closely and distantly related fungal species.

3.13 PROTEIN CODING GENES

Protein coding genes are the DNA sequences that lead to the production of different proteins required for cell functioning via transcription and translation. These genes consist of transcribed and non-transcribed regions. The region present upstream to transcribed part contains most of the signals that regulate transcription process and this region is called promoter region present at 5′ end. 3′ end contains signals for termination of transcription.

3.13.1   TUBULIN

Tubulin is a globular heterodimeric protein and is classified as alpha (α) tubulin and beta (β) tubulin. The molecular weight of tubulin is about 55 kDa and tubulins are responsible for the production of microtubules. β-tubulins are encoded by highly conserved multigene families or in some instances, by a single gene. On the basis of tubulin genes wide variety of organisms are identified and characterized. Thus, inter- and intra-specific relationship between these organisms can be successfully determined. Either β-tubulin or calmodulin sequences can be used for accurate species identification (Kövics et al., 2005; Lee et al., 2008; Irinyi et al., 2009). These sequences are easily available in public data-bases and have been used universally for their relatively high resolving power (Landvik et al., 2001).

3.13.2   TRANSLATION ELONGATION FACTORS (TEF) GENE

Elongation factors are set of proteins used in protein synthesis. They facilitate elongation in protein translation starting from the formation of first peptide bond to the last bond in ribosome. The translation elongation factor 1 gene (tef1) has been used extensively for phylogenetic and taxonomic evaluation studies (Schoch et al., 2006; Zhang et al., 2009). The advantage of using single copy genes as identification tool is that, unspecified sequence variation within a spore is certainly identified to divergence among nuclei. According to Geiser et al. (2004) partial translation elongation factor 1-alpha sequences from *Fusarium* sp, have been used as powerful single-locus identification tools.

3.13.3   ACTIN GENE

Actin gene codes for a multifunctional highly conserved protein actin, which is found in all eukaryotic cells. These genes have been used for evolutionary relationship studies among fungal species. These genes also assist in exploring additional information about genetic structure in fungi (Reeb et al., 2004).

3.14 FUNGAL DNA BARCODING

DNA barcoding is an important aid for taxonomic workflow and is not a replacement for comprehensive taxonomic analysis (Hajibabaei et al., 2007; Pečnikar and Buzan, 2014). Till date, fungal collections were identified using morphological features and only a skilled technician can make routine identifications using morphological keys. In most cases, an experienced professional taxonomist is required, whereas in some cases, experts may not be able to make correct identification of fungal specimens which are damaged or in an immature stage of development. This recent taxonomic approach solves these problems (Taylor et al., 2007; Gilmore et al., 2009; Seena et al., 2010; Eberhardt, 2012). It uses short DNA markers to identify a particular fungal taxon and the methodology used to develop DNA barcodes are straight forward. The prime advantage of this method is that, even non-specialists can obtain DNA barcodes from various individuals using tiny amounts of tissue. It goes without saying that traditional taxonomy has become less important and has been replaced with modern identification methods. Although, DNA barcodes are used to identify unknown species and to assess phylogenetic relationship of known species, the use of DNA barcoding has been the subject to extensive debate (Jin et al., 2013).

The most preferred DNA barcode region for fungi is ITS; but are not enough to delineate all the taxa. In some cases, actin gene has also been considered for developing DNA barcodes for fungal species (Aveskamp et al., 2010). Assembling the Fungal Barcode of Life (AFTOL) provided a multi-gene phylogeny of the Kingdom Fungi based on upto six genes. Some of these ribosomal polymerase B2 (rpb2) and translation elongation factor 1-α (tef1-α/EF1/TEF) genes provide good resolution at the species level for many fungal groups (Spatafora, 2005). Although, broadly useful PCR and sequencing primers has so far not been designed. ITS and translation elongation factor 1-alpha (*tef1*) based DNA barcodes have been developed recently for identification of *Trichoderma* and *Hypocrea* species (Druzhinina et al., 2005). The ITS based DNA barcodes was also found to be useful in fungal groups, such as *Zygomycetes* (Schwarz et al., 2006), dematiaceous fungi (Desnos-Ollivier et al., 2006) and *Trichophyton* species (Summerbell et al., 2007), for species level identification. FU.S.A.RIUM-ID v.1.0, a publically available sequence database of partial tef1 sequences was also developed for the identification of *Fusarium* species (Geiser et al., 2004). Although, mitochondrial cytochrome c oxidase 1 (*COX1*) gene supports *Penicillium* identification (Seifert et al., 2007), length of fungal COX1 is highly variable (1.6–22 kb). For example, the region is exceptionably variable in arbuscular mycorrhizal fungi (Glomeromycota) and hence COX1 does not resolve closely related species. In this context, Stockinger (2010) proposed the sequencing of easily amplifiable SSUmCf-LSUmBr 1500 bp fragment variants by PCR for *Glomeromycota* DNA barcoding.

3.15 APPLICATIONS OF MOLECULAR MARKERS IN FUNGI

Molecular markers are most reliable and hence extensively used in species identification. A marker allows the direct identification of the gene of interest rather the gene product and hence, it serves as a vital tool for screening fungal species. Different types of molecular markers are available for identification using linkage analysis. Molecular markers are broadly classified into two classes. Type I markers find its application in comparative mapping strategies, where, polymorphism is not an essential criterion. Type I markers represent the evolutionary conserved coding sequences (e.g., classical RFLPs and SSLPs). Type II marker, such as microsatellites, have higher polymorphism information than conventional RFLPs and they can be produced easily and in short time. Therefore, major efforts are concentrated towards the production of type II markers. Molecular markers can unravel genetic variations in both, coding and non-coding sequence regions. DNA polymorphisms that occur in and around the structural and/or regulatory sequences of a gene of physiological significance, such as hormone genes, may directly affect gene expression by changing the splicing of mRNA, stability of mRNA, rate of gene transcription or the sequence of gene product and thereby contribute to the phenotypic variations among the individuals in terms of productivity, health, disease resistance, and susceptibility. Consequently, such DNA polymorphisms, occurring in the genes, which already have the possibility to be associated or closely linked with the performance trait of importance, can be selected as molecular markers (Thorpe et al., 1994). Studies have shown that a number of single point mutations in structural genes that are inherited in a simple Mendelian manner and they are associated with quantitative traits of economic importance.

The variations occurring in non-coding sequences, such as flanking regions or intergenic regions, are used indirectly as markers for linkage analysis. Microsatellite markers, usually highly polymorphic, are currently being exploited to identify fungi. Molecular approaches provide with great range of variable characters for fungal taxonomy; they can be generated using a widely available technology. A technology that comes with an extremely well-developed bioinformatics' tools, that allows worldwide communication and comparison of results. The result produced generally correlate with reproductive barriers and physiological differences. This utility proves that molecular characters will have a primary role in determining fungal taxa. However, good taxonomy does not complete with just the recognition of a species and a Latin binomial. Species descriptions should include data from as many sources as possible, comprising of morphology, physiology and molecular data, which can be used not only as tools for identifying an isolate, but also to understand its biology (Tinker et al., 1993).

KEYWORDS

- Allele-specific associated primer
- Amplified fragment length polymorphism
- Annealing temperatures
- Arbitrary primed polymerase chain reaction
- Autoradiography
- Bacteriophages
- Calmodulin
- Codominant markers
- Comparative genetics
- DNA amplification

- **DNA amplification fingerprinting**
- **DNA fingerprinting**
- **Endophytic fungi**
- **Expressed sequence tags**
- **Externally transcribed spacer**
- **Filler bases**
- **Fungal DNA barcoding**
- **Gene mapping**
- **Gene probing**
- **Genetic identity**
- **Genetic loci**
- **Genetic mapping**
- **Genetic typing**
- **Genomic libraries**
- **Hypervariable regions**
- **Intergenic spacer**
- **Internal transcriber spacer region**
- **Marker-assisted selection**
- **Microsatellites**
- **Minisatellites**
- **Molecular phylogeny**
- **Molecular scissors**
- **Multi-gene phylogeny**
- **Nitrocellulose membrane**
- **Non-transcribed spacer**
- **Nuclear internal transcribed spacer**
- **Nuclear rDNA phylogeny**
- **Oligonucleotide fingerprints**
- **Open reading frames**
- **PCR-based markers**
- **Phylogenetic relationship**
- **Polyacrylamide gel electrophoresis**

- **Promoter region**
- **Protein coding genes**
- **Random amplified polymorphic DNA**
- **Repetitive DNA**
- **Restriction digestion**
- **Restriction endonucleases**
- **Restriction fragment length polymorphism**
- **Ribosomal RNA gene clusters**
- **Sequence-characterized amplified regions**
- **Sequence-related amplified polymorphism**
- **Short tandem repeats**
- **Silver staining**
- **Simple sequence repeats**
- **Single-strand conformation polymorphism**
- **Transcription**
- **Translation elongation factor 1 gene**
- **Tubulin**
- **Variable number tandem repeats**

REFERENCES

Aveskamp, M. M., de Gruyter, J., Woudenberg, J. H. C., Verkley, G. J. M., Crous, P. W. (2010). Highlights of the *Didymellaceae*: a polyphasic approach to characterize *Phoma* and related pleosporalean genera. *Stud Mycol.* 65, 1–60.

Bowman, B. H., Taylor, J. W., Brownlee, A. G., Lee, J., Lu, S.-D., White, T. J. (1992). Molecular evolution of the fungi: relationship of the Basidiomycetes, Ascomycetes, and Chytridiomycetes. *Mol. Biol. Evol.* 9(2), 285–296.

Brenner, E. D., Stevenson, D. W., McCombie, R. W., Katari, M. S., Rudd, S. A., Mayer, K. F. X., Palenchar, P. M., Runko, S. J., Twigg, R. W., Dai, G. W., Martienssen, R. A., Benfey, P. N., Coruzzi, G. M. (2003). Expressed sequence tag analysis in *Cycas*, the most primitive living seed plant. *Genome Biol.* 4, R78.

Brisbane, P. G., Neate, S. M., Pankhurst, C. E., Scott, N. S., Thomas, M. R. (1995). Sequence-tagged site markers to identify *Rhizoctonia solani* AG 4 or 8 infecting wheat in South Australia. *Phytopathology* 85, 1423–1427.

Bruns, T. D., White, T. J., Taylor, J. W. (1991). Fungal molecular systematics. *Annu. Rev. Ecol. Syst.* 22, 525–564.

Caldeira, A. T., Salvador, C., Pinto, F., Arteiro, J. M., Martins, M. R. (2009). MSP-PCR and RAPD molecular biomarkers to characterize *Amanita ponderosa* mushrooms. *Ann. Microbiol.* 59, 629–634.

Charter, N. W., Buck, K. W., Brasier, C. M. (1996). Multiple insertions and deletions determine the size differences between the mitochondrial DNAs of the EAN and NAN races of *Ophiostoma novo-ulmi*. *Mycological Research* 100(3), 368–372.

Choi, Y.-J., Hong, S. B., Shin, H. D. (2006). Genetic diversity within the *Albugo candida* complex (Peronosporales, Oomycota) inferred from phylogenetic analysis of ITS rDNA and cox2 mtDNA sequences. *Molecular Phylogenetics and Evolution* 40, 400–409.

Cobb, B. D., Clarkson, J. M. (1993). Detection of molecular variation in the insect pathogenic fungus *Metarhizium* using RAPD-PCR. *FEMS Microbiol.* Lett. 112(3), 319–324.

Coddington, A., Matthews, P. M., Cullis, C., Smith, K. H. (1987). Restriction digest patterns of total DNA from different races of *Fusarium oxysporum* f sp. *pisi* – an improved method for race classification. *Journal of Phytopathology* 118(1), 9–20.

Cooley, R. N. (1992). The use of RFLP analysis, electrophoretic karyotyping and PCR in studies of plant pathogenic fungi, *In Molecular Biology of Filamentous Fungi*, ed. by Stahl, U., P. Tudzynski. Verlag Chemi Weinheim, New York. pp. 13–26.

Desnos-Ollivier, M., Bretagne, S., Dromer, F., Lortholary, O., Dannaoui, E. (2006). Molecular identification of black-grain mycetoma agents. *Journal of Clinical Microbiology* 44, 3517–3523.

Druzhinina, I. S., Kopchinskiy, A. G., Komon, M., Bissett, J., Szakacs, G., Kubicek, C. P. (2005). An oligonucleotide barcode for species identification in *Trichoderma* and *Hypocrea*. *Fungal Genetics and Biology* 42, 813–828.

Eberhardt, U. (2012). Methods for DNA barcoding of fungi. *Methods Mol Biol* 858, 183–205.

Ellegren, H. (2004). Microsatellites: simple sequences with complex evolution. *Genetics* 5, 435–445.

Ferriol, M., Pico, B., Nuez, F. (2003). Genetic diversity of a germplasm collection of *Cucurbita pepo* using SRAP and AFLP markers. *Theoretical and Applied Genetics* 107, 271–282.

Gardes, M., Bruns, T. D. (1993). ITS primers with enhanced specificity for basidiomycetes; application to the identification of mycorrhizae and rusts. *Molecular Ecology* 2, 113–118.

Geisen, R. (1995). Characterization of the species *Penicillium nalgiovense* by RAPD and protein patterns and its comparison with *Penicillium chrysogenum*. *Systematic and Applied Microbiology* 18(4), 595–601.

Geiser, D. M., Jiménez-Gasco, M. M., Kang, S., Makalowska, I., Veeraraghavan, N., Ward, T. J., Zhang, N., Kuldau, G. A., O'Donnell, K. (2004). Fusarium-ID v. 1.0, A DNA sequence database for identifying *Fusarium*. *European Journal of Plant Pathology* 110, 473–479.

Giachini, A. J., Hosaka, K., Nouhra, E., Spatafora, J., Trappe, J. M. (2010). Phylogenetic relationships of the Gomphales based on nuc-25S-rDNA, mit-12S-rDNA, and mit-atp6-DNA combined sequences. *Fungal Biology* 114, 224–234.

Gilmore, S. R., Gräfenhan, T., Louis-Seize, G., Seifert, K. A. (2009). Multiple copies of cytochrome oxidase 1 in species of the fungal genus *Fusarium*. *Mol Ecol Resour* 9(S1), 90–98.

Grossman, Z., Berns, K. I., Winocour, E. (1985). Structure of simian virus 40-adeno-associated virus recombinant genomes. *J. Virol.* 56(2), 457–465.

Hajibabaei, M., Singer, G. A. C., Hebert, P. D. N., Hickey, D. A. (2007). DNA barcoding: how it complements taxonomy, molecular phylogenetics and population genetics. *Trends in Genetics* 23(4), 167–172.

Irinyi, L., Gade, A. K., Ingle, A. P., Kövics, G. J., Rai, M. K., Sándor, E. (2009). Morphology and molecular biology of *Phoma. In Current advances in molecular mycology*, ed. by Gherbawy, Y., Mach, R. L., Rai, M. K., Nova Science Publishers, Inc., New York, pp. 171–203.

Jin, Q., Huilin, H., Hu, X. M., Li, X. H., Zhu, C. D., Ho, S. Y. W., Ward, R. D., A.-Zhang, B. (2013). Quantifying species diversity with a DNA barcoding-based method: Tibetan moth species (Noctuidae) on the Qinghai-Tibetan plateau. *PLoS One* 8(5), e64428.

Kong, P., Hong, C., Richardson, P. A., Gallegly, M. E. (2003). Single-strand-conformation polymorphism of ribosomal DNA for rapid species differentiation in genus *Phytophthora. Fungal Genetics and Biology* 39(3), 238–249.

Kövics, G. J., Pandey, A. K., Rai, M. K. (2005). *Phoma saccardo* and related genera: some new perspectives in taxonomy and biotechnology. *In Biodiversity of fungi: their role in human life*, ed. by Deshmukh, S. K., Rai, M. K., Science Publishers, Inc., Enfield, USA, pp. 129–154.

Kuninaga, S., Natsuaki, T., Takeuchi, T., Yokosawa, R. (1997). Sequence variation of the rDNA ITS regions within and between anastomosis groups in *Rhizoctonia solani. Current Genetics* 32, 237–243.

Landvik, S., Eriksson, O. E., Berbee, M. L. (2001). Neolecta- a fungal dinosaur? Evidence from beta-tubulin amino acid sequences. *Mycologia* 93, 1151–1163.

Lee, R. C. H., Williams, B. A. P., Brown, A. M. V., Adamson, M. L., Keeling, P. J. (2008). α- and β-tubulin phylogenies support a close relationship between the microsporidia *Brachiola algerae* and *Antonospora locustae. J. Eukaryot. Microbiol.* 55(5), 388–392.

Li, G., Quiros, C. F. (2001). Sequence-related amplified polymorphism (SRAP), a new marker system based on a simple PCR reaction: its application to mapping and gene tagging in Brassica. *Theoretical and Applied Genetics* 103(2–3), 455–461.

Lin, Z., He, D., Zhang, X., Nie, Y., Guo, X., Feng, C., Stewart, J. M. D. (2005). Linkage map construction and mapping QTL for cotton fiber quality using SRAP, SSR and RAPD. *Plant Breeding* 124, 180–187.

Marriott, A. C., Archer, S. A., Buck, K. W. (1984). Mitochondrial DNA in *Fusarium oxysporum* is a 46.5 kilobasepair circular molecule. *Journal of General Microbiology* 130, 3001–3008.

Mavridou, A., Typas, M. A. (1998). Intraspecific polymorphisms in *Metarhizium anisopliae* var. *anisopliae* revealed by analysis of rRNA gene complex and mtDNA RFLPs. *Mycol. Res.* 102, 1233–1241.

Migheli, Q., Briatore, E., Garibaldi, A. (1998). Use of random amplified polymorphic DNA (RAPD) to identify races 1, 2, 3 and 8 of *Fusarium oxysporum* f. sp. *dianthi* in Italy. *Eur. J. Plant Pathol.* 104(1), 49–57.

Mutlu, N., Boyac, F. H., Göçmen, M., Abak, K. (2008). Development of SRAP, SRAP-RGA, RAPD and SCAR markers linked with a *Fusarium* wilt resistance gene in eggplant. *Theor. Appl. Genet.* 117, 1303–1312.

Orita, M., Iwahana, H., Kanazawa, H., Hayashi, K., Sekiya, T. (1989). Detection of polymorphisms of human DNA by gel electrophoresis as single-strand conformation polymorphisms. *Proc. Natl. Acad. Sci. USA* 86(8), 2766–2770.

Paffetti, D., Barberio, C., Casalone, E., Cavalieri, D., Fani, R., Fia, G., Mori, E., Polsinelli, M. (1995). DNA fingerprinting by random amplified polymorphic DNA and restriction fragment length polymorphism is useful for yeast typing. *Res. Microbiol.* 146(7), 587–594.

Pečnikar, Ž.F., Buzan, E. V. (2014). 20 years since the introduction of DNA barcoding: from theory to application. *Journal of Applied Genetics* 2014, 55(1), 43–52.

Reeb, V., Lutzoni, F., Roux, C. (2004). Contribution of RPB2 to multilocus phylogenetic studies of the euascomycetes (Pezizomycotina, Fungi) with special emphasis on the lichen-forming Acarosporaceae and evolution of polyspory. *Molecular Phylogenetics and Evolution* 32, 1036–1060.

Rouland-Lefevre, C., Diouf, M. N., Brauman, A., Neyra, M. (2002). Phylogenetic relationships in Termitomyces (Family Agaricaceae) based on the nucleotide sequence of ITS: A first approach to elucidate the evolutionary history of the symbiosis between fungus-growing termites and their fungi. *Molecular Phylogenetics and Evolution* 22, 423–429.

Schoch, C. L., Shoemaker, R. A., Seifert, K. A., Hambleton, S., Spatafora, J. W., Crous, P. W. (2006). A multigene phylogeny of the Dothideomycetes using four nuclear loci. *Mycologia* 98(6), 1041–1052.

Schwarz, P., Bretagne, S., Gantier, J. C., Garcia-Hermoso, D., Lortholary, O., Dromer, F., Dannaoui, E. (2006). Molecular identification of *Zygomycetes* from culture and experimentally infected tissues. *Journal of Clinical Microbiology* 44, 340–349.

Seena, S., Pascoal, C., Marvanová, L., Cássio, F. (2010). DNA barcoding of fungi: a case study using ITS sequences for identifying aquatic hyphomycete species. *Fungal Divers* 44, 77–87.

Seifert, K. A., Samson, R. A., Dewaard, J. R., Houbraken, J., Levesque, C. A., Moncalvo, J. M., Louis-Seize, G. G., Hebert, P. D. (2007). Prospects for fungus identification using CO1 DNA barcodes, with *Penicillium* as a test case. *Proceedings of the National Academy of Sciences USA* 104, 3901–3906.

Shao, Y., Xu, L., Chen, F. (2011). Genetic diversity analysis of *Monascus* strains using SRAP and ISSR markers. *Mycoscience* 52(4), 224–233.

Sharon, M., Kuninaga, S., Hyakumachi, M., Sneh, B. (2006). The advancing identification and classification of *Rhizoctonia* spp. using molecular and biotechnological methods compared with the classical anastomosis grouping. *Mycoscience* 47, 299–316.

Spatafora, J. W. (2005). Assembling the fungal tree of life (AFTOL). *Mycological Research* 109, 755–756.

Stockinger, H., Krüger, M., Schüßler, A. (2010). DNA barcoding of arbuscular mycorrhizal fungi. *New Phytologist* 187(2), 461–474.

Strommer, J., Peters, J., Zethof, J., de Keukeleire, P., Gerats, T. (2002). AFLP maps of *Petunia* hybrid; building maps when markers cluster. *Theoretical and Applied Genetics* 105, 1000–1009.

Stukenbrock, E. H., Banke, S., Zala, M., Mcdonald, B. A., Oliver, R. P. (2005). Isolation and characterization of EST-derived microsatellite loci from the fungal wheat pathogen *Phaeosphaeria nodorum*. *Molecular Ecology Notes,* doi: 10.1111/j.1471-8286.2005.01120.x.

Summerbell, R. C., Moore, M. K., Starink-Willemse, M., Iperen, A. V. (2007). ITS bar-codes for *Trichophyton tonsurans* and *T. equinum*. *Medical Mycology* 45, 193–200.

Sun, S. J., Gao, W., Lin, S. Q., Zhu, J., Xie, B. G., Lin, Z. B. (2006). Analysis of genetic diversity in *Ganoderma* population with a novel molecular marker SRAP. *Appl Microbiol Biotechnol*. 72, 537–543.

Tanabe, Y., O'Donnell, K., Saikawa, M., Sugiyama, J. (2000). Molecular phylogeny of para-sitic Zygomycota (Dimargaritales, Zoopagales) based on nuclear small subunit ribo-somal DNA sequences. *Molecular Phylogenetics and Evolution* 16, 253–262.

Taylor, J. W., Turner, E., Pringle, A., Dettman, J., Johannesson, H. (2007). Fungal species: thoughts on their recognition, maintenance and selection. *In Fungi in the environment*, ed. by Gadd, G. M., Watkinson, S. C., Dyer, P. S., Cambridge University Press, Cambridge, pp. 313–339.

Thorpe, R. S., McGregor, D. P., Cumming, A. M., Jordan, W. C. (1994). DNA evolution and colonization sequence of island lizards in relation to geological history: mtDNA RFLP, cytochrome B, cytochrome oxidase, 12S rRNA sequence, and nuclear RAPD analysis. *Evolution* 48(2), 230–240.

Tinker, N. A., Fortin, M. G., Mather, D. E. (1993). Random amplified polymorphic DNA and pedigree relationships in spring barley. *Theoretical and Applied Genetics* 85(8), 976–984.

Toda, T., Nasu, H., Kageyama, K., Hyakumachi, M. (1998). Genetic identification of the web-blight fungus (*Rhizoctonia solani* AG 1) obtained from European pear using RFLP pf rDNA-ITS and RAPD analysis. *Res. Bull. Fac Agric Gifu Univ*. 63, 1–9.

Varga, J., Kevei, F., Fekete, C., Coenen, A., Kozakiewicz, Z., Croft, J. H. (1993). Restric-tion fragment length polymorphisms in the mitochondrial DNAs of the *Aspergillus niger* aggregate. *Mycological Research* 97, 1207–1212.

Voort, V. J. N. A. M. R., Zandvoort, V. P., Eck, V. H. J., Folkertsma, R. T., Hutten, R. C. B., Draaistra, J., Gommers, F. J., Jacobsen, E., Helder, J., Bakker, J. (1997). Use of allele specificity of comigrating AFLP markers to align genetic maps from different potato genotypes. *Molecular and General Genetics* 255, 438–447.

Vos, P., Hogers, R., Bleeker, M., Reijans, M., Lee, V. T., Hornes, M., Frijters, A., Pot, J., Peleman, J., Kuiper, M., Zabeau, M. (1995). AFLP: a new technique for DNA finger-printing. *Nucleic Acids Research* 23, 4407–4414.

Wang, Z., Binder, M., Schoch, C. L., Johnston, P. R., Spatafora, J. W., Hibbett, D. S. (2006). Evolution of helotialean fungi (Leotiomycetes, Pezizomycotina): a nuclear rDNA phylogeny. *Molecular Phylogenetics and Evolution* 41, 295–312.

Welsh, J., McClelland, M. (1991). Genomic fingerprinting using arbitrarily primed PCR and a matrix of pairwise combinations of primers. *Nucleic Acids Res*. 19(19), 5275–5279.

Zabeau, M., Vos, P. (1993). Selective restriction fragment amplification: a general method for DNA fingerprinting. *European Patent Publication* 92402629 (Publication No. EP0534858A1).

Zeng, Q.-Y., X.-Wang, R., Blomquist, G. (2003). Development of mitochondrial SSU rDNA-based oligonucleotide probes for specific detection of common airborne fungi. *Molecular and Cellular Probes* 17, 281–288.

Zhang, Y., Schoch, C. L., Fournier, J., Crous, P. W., Gruyter, J., Woudenberg, J. H. C., Hirayama, K., Tanaka, K., Pointing, S. B., Spatafora, J. W., Hyde, K. D. (2009). Multi-locus phylogeny of Pleosporales: a taxonomic, ecological and evolutionary re-evaluation. *Stud. Mycol*. 64(S5), 85–102.

CHAPTER 4

GENETIC IMPROVEMENT OF INDUSTRIALLY IMPORTANT FUNGAL STRAINS

CONTENTS

4.1   INTRODUCTION

The unique metabolic pathways of fungal molecules serve as a rich source of many industrially useful active components. Fungal production of organic acids and other metabolites is an excellent approach for obtaining different commercially useful products. Production has been started decades ago and titers have been improved by classical approaches like mutation and screening, selection techniques as well as by metabolic and genetic engineering. Ancient Babylonians and Sumerians were brewing as early as 6000 BC and reliefs on tombs dating from 2400 BC documented beer making in Egypt. Most useful industrial fungi are *Aspergillus niger, A. terreus, A. oryzae, A. sojae, Candida utilis, Cladosporium resinae, Cephalosporium acremonium, Fusarium moniliforme, Gibberella fujikoroi, Lecanicillium* sp., *Morchella esculenta, Paecilomyces* sp., *Penicillium chrysogenum, P. griseofulvum, P. notatum, P. roqueforti, P. camemberti, Rhizopus nigricans, Saccharomyces cerevisiae, S. lipolytica, S. rouxi, Tolypocladium inflatum, Trichoderma harziannum, T. ressei* and *T. viridae*.

4.2   FUNGAL APPLICATIONS IN INDUSTRIAL SECTORS

Fungi are most efficient producers of large variety of enzymes and chemicals. Their ability to grow at low pH values and high salt concentrations and to utilize C5 and C6 sugars as substrates makes them favorable for industrial application. They can withstand extreme high and low temperatures also. Their mode of reproduction so as to give high number of multiples within short term is another advantage. They require little space to work and it is easy to maintain the consistency of the reaction when compared to the other organisms. Extracellular liberation of their products makes it easier to extract and purify. All the commercially valuable fine and bulk products produced by them can be categorized into their primary as well as secondary metabolites.

Primary metabolite of an organism is a key component in maintaining normal physiological process and is involved in growth, development and reproduction of the organism. They include alcohols such as ethanol, lactic acid, and certain supplementary amino acids such as L-glutamate. Among them, alcohol which includes beer and wine is the most common primary metabolites used for large-scale production. The primary metabolite citric acid, which is used as ingredients in food production, pharmaceutical and cosmetic industries is produced from *Aspergillus niger*. Fungi also produce a wide range of natural products often called secondary metabolites. Secondary metabolism is commonly associated with sporulation processes in microorganisms (Stone and Williams, 1992). Though they do not have any role in their own growth and development, they have various industrial applications. They include toxins, plant hormone such as gibberellins produced by *Gibberella fujikuroi*, alkaloids and antibiotics. Secondary metabolites associated with sporulation can be placed into three broad categories: (i) metabolites that activate sporulation such as linoleic acid-derived compounds produced by *A. nidulans* (Calvo et al., 2001), (ii) pigments required for sporulation structures such as melanins required for the formation or integrity of both sexual and asexual spores and overwintering bodies (Kawamura et al., 1999), and (iii) toxic metabolites secreted by growing colonies at the approximate time of sporulation like biosynthesis of some deleterious natural products like mycotoxins (Trail et al., 1995; Hicks et al., 1997). Apart from the in born production of secondary metabolites, they can be induced by giving them inappropriate growth environment like nutrient deficiency, since they produce secondary metabolites as defense mechanism.

The primary and secondary metabolic products of fungi are being used in a range of industries such as fuel production, pharmaceuticals, enzymes, hormones, baking, brewing, biofertilizers and biomining industries. The major application of the fungal metabolites especially as antibiotic is in the pharmaceutical field. They produce antibiotics, immunosuppressant and statin compounds such as rosuvastatin and levostatin, which are important in controlling cholesterol. Fungi produce organic acids during carbohydrate metabolism. They are produced in large scale due to high demand. The organic acids formed are used in medical as well as in research field. Organic acids such as lactic acid and citric acid are extensively useful in food industry. They are also used in preparation of dye mordant, cosmetics and paints. More than 260 enzyme products have been commercialized by members of the Association of Manufactures and Formulators of Enzyme Products (AMFEP). Nearly 60% of these commercial enzymes originate from fungi and they are also produced in fungal host organism. Enzymes like digestin and amylase are produced with the help of *Aspergillus flavus* and *Aspergillus niger* respectively.

Currently, biofuel is the most promising alternative for the energy crisis and high market price of the fuel. Fungi can efficiently convert biomass into energy in the form of alcohol. Any wastes which are rich in carbohydrate can be converted to energy. It can be prepared by using cassava, yams, sugar beets, sugar beet molasses, spent mushroom substrates, cane juice, maize, corn, sorghum and agri-wastes. *Saccharomyces cerevisiae* is used widely for the ethanol production using various raw materials as substrate. *Mucor indicus*, a saprophytic zygomycetes has been reported as ethanol producing fungi recently. Fungi establish the light and spongy products by making the dough to rise. *Saccharomyces* sp. helps to ferment the carbohydrate to form carbon dioxide and alcohol. Elastic extension of gluten, which is a protein in the flour, is caused by the bubbles of carbon dioxide produced by the fermentation process. Hence, the length of leavening and gluten present in the flour are the texture and flavor determinants. Soy sauce is a widely used fermented product of *C. lipolytica* and *A. oryzae*. Microbial proteins which are produced by mass culture of the microbes are widely used in different zones as human food and animal feed. They play a remarkable role in resolving world food shortages as they are protein rich supplement. *Saccharomyces cerevisiae*, *S. lipolytica*, *Aspergillus oryzae*, *Candida utilis*, *C. moniliforme* and *Morchella esculenta* are used as single cell protein (SCP). Fungi like *Saccharomyces cerevisiae* are extensively being used in alcohol production. In the process of sugar hydrolysis to pyruvic acid by Embden-Myerhof-Parnas pathway which extends to the conversion of acetaldehyde to ethanol. This exothermic reaction can yield upto 50% alcohol, by weight of sugar. Ales and wine is prepared using *S. cerevisiae*, whereas, *S. carlsbergensis* and *S. uvarum* are the fungi used for lager and cider production respectively. *S. sakeis* are being used for saké, a rice wine preparation. The byproduct of fermentation, such as lactate and glycerol, are also industrially valuable compounds.

Mushrooms belong to the class Ascomycetes and Basidomycetes and are known as 'higher fungi.' They serve as rich source of proteins and other nutrients. Their ability to grow on waste materials which are rich in carbohydrate makes them attractive for agriwaste management. *Amanita vaginata*, *Cantharellus cibarius*, *Naematoloma sublateritium*, *Hericium erinaceus*, *Heterobasidium annostum*, *Morganella pyriformis*, *Entoloma abortivum*, *Lepista nuda*, *Grifola frondosa*, *Pleurotus ostreatus*, *P. porrigens* and *Panellus serotinus* are categorized as edible, whereas, mushrooms such as *Flammulina velutipes*, is said to be consumed with precaution. Some mushrooms are found to be highly poisonous. They are being used for the production of various industrially important organic acids and antibiotics. Antioxidant, antimicrobial, anticancer, cholesterol lowering and immunostimulatory effects have been reported in some species of mushrooms (Anderson, 1992; Mau et al., 2004). These properties of mushrooms have been attributed to the presence of bioactive compounds in mushrooms. Some of these biologically active substances are glycolipids, compounds derived from shikimic acid, aromatic phenols, fatty acid derivatives, polyacetylamine, polyketides, nucleosides, sesquiterpenes and many other substances of different origins (Lorenzen and Anke, 1998; Wasser and Weis,

1999). The sources of these bioactive compounds include fruiting body, mycelia, cultivation broth, submerged cultivation mycelia and fermentation derivatives (Fiore and Kakkar, 2003; Yoon et al., 2003; Kwok et al., 2005; Yamamoto et al., 2005; Sugimoto et al., 2007; Wu et al., 2010; Wong et al., 2011). *Amanita virosa* and *Amanita verna* are example for fatal mushrooms. It is reported that microbial fibrinolytic enzymes, chiefly those from food-grade microorganisms, have prospective ability to be developed as functional food additives and drugs to prevent or cure thrombotic disease (Peng et al., 2005).

4.3 NOVEL STRATEGIES FOR FUNGAL STRAIN IMPROVEMENT

In order to use fungal strain in biotechnological processes, strain improvement has to be setup to generate a stable and defined genetic background. Improvement of strain by hybridization, rare mating, spheroplast fusion, cytoduction and single chromosome transfer may be seen from 1970's. Protoplast fusion was used to improve the characteristics of penicillin-producing strain of *Penicillium chrysogenum* that showed poor sporulation and poor seed growth. To overcome this problem, backcrossing was made among low-producing strain with high producing strain that yielded strains with better sporulation and better growth in seed medium. From 1980, the application of genetic recombination for the production of important fungal products such as antibiotics was heightened. Genetic manipulation has successfully been employed to improve the ability of biocontrol agents. In order to achieve this goal, the researchers have attempted various methods like to enhance antifungal metabolites productivity in fungal strains, improving antagonistic potential of biocontrol agents, controlling a broad spectrum of phytopathogens, increasing their competitiveness potential and developing tolerant strains to stress conditions (Harman et al., 2004; Wafaa and Mohammed, 2007). During strain improvement, several strategies will affect the production level due to heterologous genes and the production can be limited at any level during transcription, translation, secretion and extracellular degradation (Punt et al., 1994; Archer and Peberdy, 1997).

Introduction of multicopy genes was another strategy developed for strain improvement. The transcriptional level can be improved by increasing the copy number. Integration of multiple copies of mycoparasitism related gene, *prb 1* into the genome of *Trichoderma atroviridae* increased its biocontrol activity (Flores et al., 1997). But sometimes, gene silencing at transcriptional and post-transcriptional level of duplicated gene copies will be observed in *Neurospora crassa, Aspergillus nidulans* and *Aspergillus flavus* (Cogoni, 2001; Clutterbuck, 2004; Schmidt, 2004; Hammond and Keller, 2005). For gene expression, there are number of deciding factors like specificity of an integration site have to be well placed in the strain improvement technology rather than random insertion. Selection of suitable strong promoter like *pcbh1* and *ech 42* is another criterion to be considered in transgenic expression modification. Understanding of the optimal codons used by an expression host allows the recognition of codons present in the desired recombinant genes that may limit the expression level of their encoded protein (Koushki et al., 2011). Fungi produce large amount of extracellular protease, which will degrade the required protein products. Therefore, it is often essential to knock out non proteases gene from strain to produce desired protein and identify genes involved in regulating extracellular protease secretion, since knocking out all native proteases would likely be lethal (Katz et al., 2000).

Lovastatin titers from *Aspergillus terreus* could be improved by increasing dosage of lovastatin biosynthetic genes and its regulatory genes are *lovF, creA, fadA, ganA, gnaI, gna3* and *gpa* for secondary metabolism. The *lovF::ble* transcriptional fusion protein was served as a reporter-based system to select improved mutants, whereas, *ble* gene encoding resistance to phleomycin (Askenazi et al., 2003). Recombinant hybrid strains, Kel+Ben(R) DBP+Lin+ with improved dehalogenase activity that shows superior degradation quality for DDT were raised by Mitra et al. (2001) using parasexual hybridization methodology of two such complementary isolates, viz. isolate 1(P-1) and 4(P-2), where they are showing highest complementation and are compatible for hyphal fusion inducing heterokaryosis which are genetically characterized as Kel+Ben(R) DBP-Lin- and Kel-Ben(r) DBP+Lin+ respectively. Confirmation of the recombination was done by polypeptide band analysis of DDD induced exo-proteins from culture filtrate using SDS-Polyacrylamide Gel Electrophoresis (PAGE) and RAPD (Random Amplified Polymorphic DNA) of genomic DNA using PCR (Polymerase Chain Reaction) technique. SDS-PAGE showed combination of DDD induced polypeptide bands characteristic of both parents in recombinants or hybrids. Parent specific bands in the recombinant strains confirming gene transformation could be detected using PCR study.

The low frequency of recombination resulted in the ignorance of genetic recombination in industry. Nevertheless, use of protoplast fusion changed the situation markedly. The frequencies of recombination have increased to even greater than 10^{-1} in some cases and strain improvement programs routinely include protoplast fusion between different mutant lines (Ryu et al., 1983). Strain improvement for tofty productivity is highly necessitated. Till date, remarkable improvements in the productivity of many primary and secondary metabolites as well as protein biopharmaceuticals and enzymes could be brought forward by the amalgamation of complementary technologies such as classical and novel methods. Improvement of the suitable strain with appropriate methodology in order to achieve implementation of the higher productivity may require detailed analysis over the work done on the field. This will help in the development of the existing methodologies and emergence of new technologies to be par with the discovery of new fungal bioactive compounds.

KEYWORDS

- Alcohol production
- Amylase
- Antifungal metabolites
- *Aspergillus flavus*
- *Aspergillus nidulans*
- *Aspergillus niger*
- *Aspergillus oryzae*
- Backcrossing
- Baking
- Biocontrol agents

- **Biofuel**
- **Biomining**
- **Brewing**
- *Candida utilis*
- **Carbohydrate metabolism**
- *Cephalosporium acremonium*
- *Cladosporium resinae*
- **Cytoduction**
- **Digestin**
- **Embden-Myerhof-Parnas pathway**
- **Extracellular protease**
- *Flammulina velutipes*
- **Fungal biomolecules**
- *Fusarium moniliforme*
- **Gene expression**
- **Gene silencing**
- **Genetic engineering**
- **Genetic manipulation**
- *Gibberella fujikoroi*
- **Heterokaryosis**
- **Higher fungi**
- **Immunosuppressant**
- **Metabolic engineering**
- **Microbial proteins**
- *Morchella esculenta*
- *Mucor indicus*
- **Mycotoxins**
- *Neurospora crassa*
- **Organic acids**
- **Parasexual hybridization**
- *Penicillium chrysogenum*

- **Phytopathogens**
- **Polyacrylamide gel electrophoresis**
- **Protoplast fusion**
- **Random insertion**
- **Rare mating**
- **Recombinant hybrid strains**
- *Rhizopus nigricans*
- *Saccharomyces cerevisiae*
- **Secondary metabolites**
- **Single cell protein**
- **Spheroplast fusion**
- **Strain improvement**
- *Tolypocladium inflatum*
- *Trichoderma atroviridae*
- *Trichoderma harziannum*

REFERENCES

Anderson, J. B., Stasovski, E. (1992). Molecular phylogeny of Northern Hemisphere species of *Armillaria*. *Mycologia* 84, 505–516.

Archer, D. B., Peberty, J. F. (1997). The molecular biology of secreted enzyme production by fungi. *Crit. Rev. Biotechnol.* 17, 273–306.

Askenazi, M., Driggers, E. M., Holtzman, D. A., Norman, T. C., Iverson, S., Zimmer, D. P. (2003). Integrating transcriptional and metabolite profiles to direct the engineering of lovastatin-producing fungal strains. *Nat. Biotechnol.* 21, 150–156.

Calvo, A. M., Gardner, H. W., Keller, N. P. (2001). Genetic connection between fatty acid metabolism and sporulation in *Aspergillus nidulans*. *J. Biol. Chem.* 276, 25766–25774.

Clutterbuck, A. J. (2004). MATE transposable elements in *A. nidulans*: evidence of repeat-induced point mutation. *Fungal Genetics and Biology* 41(3), 308–316.

Cogoni, C. (2001). Homology-dependent gene silencing mechanisms in fungi. *Ann. Rev. Microbiol.* 55, 381–406.

Fiore, M. M., Kakkar, V. V. (2003). Platelet factor 4 neutralizes heparan sulfate-enhanced antithrombin inactivation of factor Xa by preventing interaction(s) of enzyme with polysaccharide. *Biochemical and Biophysical Research Communications* 311(1), 71–76.

Flores, A., Chet, I., Estrella, A. H. (1997). Improved biocontrol activity of the proteinase encoding gene prb1. *Curr. Genet.* 31, 30–37.

Hammond, T. M., Keller, N. P. (2005). RNA silencing in *A. nidulans* independent of RNA dependent RNA polymerases. *Genetics* 169(2), 607–617.

Harman, G. E., Howell, C. R., Viterbo, A., Chet, I., Lorito, M. (2004). *Trichoderma* species opportunistic, avirulent plant symbionts. *Nature Rev.* 2, 43–56.

Hicks, J. K., Yu, J. H., Keller, N. P., Adams, T. H. (1997). Aspergillus sporulation and mycotoxin production both require inactivation of the FadA G alpha protein-dependent signaling pathway. *EMBO J.* 16(16), 4916–4923.

Katz, M. E., Masoumi, A., Burrows, S. R., Shirtliff, C. G., Cheetham, B. F. (2000). The *A. nidulans* xprF gene encodes a hexokinase-like protein involved in the regulation of extracellular proteases. *Genetics* 156(4), 1559–1571.

Kawamura, C., Tsujimoto, T., Tsuge, T. (1999). Targeted disruption of a melanin biosynthesis gene affects conidial development, UV tolerance in the Japanese pear pathotype of *Alternaria alternata*. *Mol Plant Microbe Interact.* 12(1), 59–63.

Koushki, M. M., Rouhani, H., Farsi, M. (2011). Genetic manipulation of fungal strains for the improvement of heterologous gene expression (a mini-review). *Afr. J. Biotechnol.* 10(41), 7939–7948.

Kwok, Y., Ng, K. F., Li, C. C., Lam, C. C., Man, R. Y. (2005). A prospective, randomized, double-blind, placebo-controlled study of the platelet and global hemostatic effects of *Ganoderma lucidum* (Ling-Zhi) in healthy volunteers. *Anesthesia and Analgesia* 101(2), 423–426.

Lorenzen, K., Anke, T. (1998). Basidiomycetes as a source for new bioactive natural products. *Current Organic Chemistry* 2, 329–364.

Mau, J. L., Chang, C. N., Huang, S. J., Chen, C. C. (2004). Antioxidant properties of methanolic extracts from *Grifola frondosa*, *Morchella esculenta* and *Termitomyces albuminosus* mycelia. *Food Chemistry* 87, 111–118.

Mitra, J., Mukherjee, P. K., Kale, S. P., Murthy, N. B. (2001). Bioremediation of DDT in soil by genetically improved strains of soil fungus *Fusarium solani*. *Biodegradation* 12(4), 235–245.

Peng, Y., Yang, X., Zhang, Y. (2005). Microbial fibrinolytic enzymes: an overview of source, production, properties, and thrombolytic activity *in vivo*. *Applied Microbiology and Biotechnology* 69(2), 126–132.

Punt, P. J., Veldhuisen, G., van den Hondel, C. A. M. J. J. (1994). Protein targeting and secretion in filamentous fungi. *Anton Leeuw Int. J.* 65, 211–216.

Ryu, D. D. Y., Kim, K. S., Cho, N. Y., Pai, H. S. (1983). Genetic recombination in *Micromonospora rosaria* by protoplast fusion. *Appl Environ Microbiol.* 45, 1854–1858.

Schmidt, F. R. (2004). RNA interference detected 20 years ago? *Nat. Biotechnol.* 22(3), 267–268.

Stone, M. J., Williams, D. H. (1992). Review on the evolution of functional secondary metabolites (natural products). *Mol Microbiol.* 6(1), 29–34.

Sugimoto, S., Fujii, T., Morimiya, T., Johdo, O., Nakamura, T. (2007). The fibrinolytic activity of a novel protease derived from a tempeh producing fungus, *Fusarium* sp BLB. *Bioscience, Biotechnology and Biochemistry* 71(9), 2184–2189.

Trail, F., Mahanti, N., Linz, J. (1995). Review molecular biology of aflatoxin biosynthesis. *Microbiology* 141(4), 755–765.

Verdoes, J. C., Punt, P. J., van den Hondel, C. A. M. J. J. (1995). Molecular genetic strain improvement for the overproduction of fungal proteins by filamentous fungi. *Applied Microbiology and Biotechnology* 43(2), 195–205.

Wafaa, H. M., Mohamed, H. A. A. (2007). Biotechnological aspects of microorganisms used in plant biological control. *World J. Agric. Sci.* 3, 771–776.

Wasser, S. P., Weis, A. L. (1999). Therapeutic effects of substances occurring in higher basidiomycetes mushrooms a modern perspective. *Critical Reviews in Immunology* 19(1), 65–96.

Wong, K. H., Lai, C. K. M., Cheung, P. C. K. (2011). Immunomodulatory activities of mushroom sclerotial polysaccharides. *Food Hydrocolloids* 25(2), 150–158.

Wu, D. M., Duan, W. Q., Liu, Y., Cen, Y. (2010). Anti-inflammatory effect of the polysaccharides of golden needle mushroom in burned rats. *International Journal of Biological Macromolecules* 46(1), 100–103.

Yamamoto, J., Yamada, K., Naemura, A., Yamashita, T., Arai, R. (2005). Testing various herbs for antithrombotic effect. *Nutrition* 21(5), 580–587.

Yoon, S. J., Yu, M. A., Pyun, Y. R., Hwang, J. K., Chu, D. C., Juneja, L. R., Mourao, P. A. S. (2003). The nontoxic mushroom *Auricularia auricula* contains a polysaccharide with anticoagulant activity mediated by antithrombin. *Thrombosis Research* 112(3), 151–158.

CHAPTER 5

FUNGAL STRAIN IMPROVEMENT THROUGH RECOMBINANT DNA TECHNOLOGY

CONTENTS

5.1   INTRODUCTION

Fungi such as yeast and filamentous fungi are the best-known eukaryotic model organism capable of post-transitional modification and secreting industrially important products (Nevalainen et al., 2005; Sharma et al., 2009). Fungi are exploited for the vast production of secondary metabolites, enzymes and antibiotics (Dufossé, 2014). Some of the properties that make yeast particularly suitable for biological studies include rapid growth, dispersed cells, the ease of replica plating and mutant isolation, well-defined genetic system, and most importantly, a highly versatile DNA transformation system (Schneiter, 2004). Strain improvement of fungi relies on random mutagenesis or on classical breeding and genetic crossing of two strains, followed by screening for mutants and progeny exhibiting enhanced properties of interest. Classical approaches such as the use of mutagenesis and selection (random screening) and recombination, is still considered to be very important in strain improvement. Considering the fact that classical approach is based on trial and error, strain development searches for new strains that give better results and more products that are capable of further development and are convenient in use. Biotechnological approaches have paved the way for bioprocess technology through newer approaches of strain improvement with retaining the impact of classical approaches. Genomics and proteomics have revolutionized the bioprocess industry with newer strain improvement techniques and bringing genetic variation and desirable characters to the microbes have resulted in huge profits of the bioprocess industries. Thus, a set of immensely powerful experimental and modeling techniques have become available in the last few decades that have enabled us to change the genome of the fungi and their products (secondary metabolites).

5.2   MUTATION

Parekh et al. (2000) stated that conventionally, strain improvement could be achieved through mutation, selection or genetic recombination. Strains were improved through adaptation, mutagenesis or gene cloning. Improvement of the microbial strain offers the greatest opportunity for cost reduction without significant outlay. Overproduction of primary and secondary metabolites is a complex process and the successful development of improved strains requires knowledge of physiology, pathway regulation and control, and the design of creative screening procedures. In addition, it also requires the mastery of the fermentation process for each new strain, as well as sound engineering know-how with media optimization and the fine-tuning of process conditions. Flickweert (1999) described the importance of classical methods and genetic engineering in strain improvement (Table 5.1).

Modifying a strain by mutating its nucleus results in high yield and this technique is followed from decades. Selection and further reproduction of the higher yielding strains over many generations can raise yields

TABLE 5.1 Industrially Important Fungi and Their Genetically Improved Strains through Mutation Techniques

Fungi	Starting strain	Mutated strain	Type of mutation	End products	References
Acremonium cellulolyticus	C-1	CF-2612	Random mutagenesis, UV, NTG	Cellulase	Fang et al. (2009)
Aspergillus niger	A	A2	UV radiation	Citric acid	Vasanthabharathi et al. (2013)
	MBL-3	MBL-3 (60 Gy)	γ-radiation	Lipase	Iftikhar et al. (2010)
	CFTRI 1105	II N 31	UV and nitrous acid mutagenesis	Asperenone	Chidananda et al. (2008)
Candida magnoliae	NCIM 3470	R23	UV and chemical mutagenesis	Erythritol	Laxman et al. (2011)
Fusarium maire	Y1117	K178	UV radiation and Diethyl sulfate (DES)	Paclitaxel (taxol)	Xu et al. (2006)
Humicola insolens	TAS-13	TAS13, UV-4, NG-5	N-methyl-N-nitro-N-nitrosoguanidine (MNNG), nitrous acid (HNO₂), ethyl methyl sulphonate (EMS)	cellulase	Javed et al. (2011)
Penicillium atrovenetum	MBL-4	MBL-4 (80 Gy)	γ-radiation	Lipase	Iftikhar et al. (2010)
Penicillium janthinellum		EU2D-21, EU1, EMS-UV-8	UV irradiation and EMS	Cellulose	Adsul et al. (2007)
Penicillium occitanis	Pol6	Mutant Pol6	NTG, EMS and UV	Cellulase and pectinase	Jain et al. (1990)

TABLE 5.1 Continued

Fungi	Starting strain	Mutated strain	Type of mutation	End products	References
Penicillium sp. EZ-ZH190	EZ-ZH190	EZ-ZH390	Heat treatment and γ irradiation	Tannase	Zakipour-Molkabadi et al. (2013)
Rhizopus microspores	MBL-5	(MBL-5 at 140 Gy)	γ radiation	Lipase	Iftikhar et al. (2010)
Rhizopus oryzae	R3017	R1021	UV irradiation, Diethyl sulfate (DES) Co60 irradiation	L(+)-lactic acid	Bai et al. (2004)
Saccharopolyspora erythraea	MTCC 1103	Mutant	UV radiation	Erythromycin	Devi et al. (2013)
Trichoderma reesei	KCTC 6950 k	mutant strain T-2 (MT-2)	Proton beam irradiation	Cellulase, β-glucosidase	Jung et al. (2012)
	Rut C-30	mutant NU-6	NTG followed by UV irradiation	Cellulose, xylanase	Jun (2009)

by 4000-fold or more (Peberdy, 1985). Isolation of an organism producing the chemical of interest using selection procedures and improvement of production yields via mutagenesis of the nucleus. Mutagenesis is the source of all genetic variations, but no single mutagenic treatment will give all possible types of mutation. Any chemical or physical agent that increases mutagenesis is referred to as a mutagen. Mutagens introduce some chemical change to DNA such as altering bases or breaking the sugar-phosphate backbone. Mutation rates can be increased by conditions that damage DNA. Mutagenesis induced by exposure to damage is defined as induced mutagenesis. A damaged base or segment of DNA is not actually a mutation; it is instead referred to as a premutational lesion. Mutagens can be broadly classified in to physical, chemical and biological mutagens. Physical mutagens such as UV, X-ray, gamma radiation, absorption of high energy ionizing radiation such as x-rays and γ-rays causes the target molecules to loose electrons. These electrons cause extensive chemical alterations to DNA, including strand breaks and base and sugar destruction eventually causing cell division blockage and cell death. Non-ionizing radiation causes molecular vibrations or promotion of electrons to higher energy levels within the target molecules. This can lead to the formation of new chemical bonds. The most important form of physical mutation causing DNA damage is UV light, which produces pyrimidine dimers from adjacent pyrimidine bases which prevent normal replication and transcription. UV radiation is considered to be most convenient for use and it has been extensively used for strain improvement.

Chemical mutagens are defined as those compounds that increase the frequency of some types of mutations. Many natural, synthetic, organic and inorganic chemicals can react with DNA changing its structures and properties. Chemical mutagens such as alkylators include ethyl methane sulfonate (EMS), methyl methane sulfonate (MMS), diethylsulfate (DES), and nitrosoguanidine (NTG, NG, MNNG). Nitrous acid is another chemical mutagen that belongs to deaminating agent class and causes oxidative deamination of particular bases. It converts adenine to hypoxanthine (which now pairs with C), cytosine to uracil (which now pairs with A) and finally guanine to xanthine and most recently, the production of cellulase from the mutant strain of *Trichoderma viride* using a new mutation technique, atmospheric pressure non-equilibrium discharge plasma (APNEDP)

(Xu et al., 2011). Biological mutagens include transposable elements, which make the genes dysfunction, Ty1 Yeast (LTR-type). The optimum concentration of mutagen is that which gives the highest proportion of desirable mutants in the surviving population. Hopwood et al. (1985) suggested that 99.9% kill is best suited for strain improvement programs as the fewer survivors in the treated sample would have undergone repeated or multiple mutations. This may lead to the enhancement in the productivity of the metabolite (Table 5.1).

5.3   GENETIC RECOMBINATION

Genetic recombination is a process which generates new combination of alleles by exchange of genetic information that was originally present in different individuals. In the early years of strain improvement era, the application of rDNA technology was fairly limited when compared with the use of mutation techniques for the strain improvement of industrially important fungi (Nevoigt, 2008). But today, rDNA technology is the most preferred method for strain improvement. Both parasexual and sexual modes of reproduction are employed in the strain improvement of industrially important fungi (Adrio and Demain, 2010). Parasexual cycle also called parasexuality is a nonsexual mode of transferring genetic materials without undergoing meiosis or the development of sexual structures. This unique process is seen in fungi and single-celled organisms (Bradley, 1962; Becker and Castro-Prado, 2006). This cycle is initiated by hyphae fusion during which nuclei and other cytoplasmic contents occupy the same cell. Fusion of dissimilar nuclei results in the formation of diploid nucleus, which is assumed to be unstable. These diploid nuclei can produce recombinant segregants by mitotic crossing over and haploidization without meiosis. Mitotic crossing over results in exchange of chromosomal genes (Virgin et al., 2001). Haploidization might cause production of aneuploid and haploid cells due to mitotic non-disjunctions, which reassort the chromosomes (Tolmsoff, 1983). The recombinant haploid nuclei produced appear among vegetative cells differing genetically from those of their parent mycelium. Parasexuality gives the organism an opportunity to introduce new genotypes into their offspring as seen in sexual cycle, but it lacks coordination and meiotic division and is exclusively mitotic. Parasexual cycle is vital for those fungi, which cannot reproduce by sexual mode. Parasexuality results in somatic variation in the vegetative phase of fungal life cycles. Though somatic variation is found in fungi reproducing sexually, but the variations introduced are additional and more significant. The possibility to mate fungi in *in vitro* conditions is a valuable tool for genetic analysis and for classical strain development (Valent and Chumley, 1991).

5.4 RECOMBINANT DNA TECHNOLOGY FOR FUNGAL IMPROVEMENT

Fungi are eukaryotic organism widely used in industry as production hosts for its ability to survive at low pH, high salt concentrations and their ability to use simple sugars as substrates (Ryu et al., 1983). Fungi are generally filamentous, thereby, increasing culture viscosity and thus, reducing oxygen transfer to the fungal cultures (Nielsen, 2001). Some of the industrially important products include enzymes, chemicals, antibiotics and celluloses (Table 5.2). Fungi generally produce protease abundantly that may degrade the protein of interest (Meyer et al., 2008). Modification of fungal strain for higher production of interested products and to maintain them at required temperature and pH is called strain improvement. rDNA technology is the most effective method employed in fungal strain improvement (Eggeling et al., 1996). Genetic engineering is revolutionizing the science of strain improvement. Recombinant DNA is a process of combining a piece of DNA with another strand of DNA or by combining two or more different strands of DNA from any species. The DNA obtained by combining DNA from two or more species is called as chimeric DNA (Burg et al., 1998). Genes that code for specific desirable traits may be derived from virtually any living organism (plant, animal, microbe or virus) on earth. In addition, DNA sequences that do not occur anywhere in nature may be created by the chemical synthesis of DNA and incorporated into recombinant molecules. Today, using recombinant DNA technology and synthetic DNA, literally any DNA sequence may be created and introduced into any form of life (Suenaga et al., 2001).

TABLE 5.2   Industrially Important Pharmaceutical Products and Organic Acids from Fungi

Secondary metabolites	Fungi	References
Pharmaceuticals		
Cephalosporin (antibiotic)	*Cephalosporium acremonium*	Hinnen and Nuesch (1976); Paul et al. (1989)
Penicillin (antibiotic)	*Penicillium chrysogenum, P. notatum*	Dayalan et al. (2011)
Griseofulvin (antibiotic)	*P. griseofulvum*	Brian (1951); Dasu et al. (2003)
Cyclosporin A (immunosuppressant)	*Trichoderma polysporum, Cylindrocarpon lucidium*	Dreyfuss et al. (1976)
Gliotoxins (immunosuppressant)	*Aspergillus fumigatus*	Sutton et al. (1994); Kupfahl et al. (2008)
Ergot alkaloids	*Claviceps purpurea, C. fusiformis, C. paspali*	Banks et al. (1974); Tudzynski et al. (2001); Wallwey and Li (2011)
Lovastatin (statin)	*A. terreus*	López et al. (2003)
Squalestatin (statin)	*Phoma* sp.	Schuemann and Hertweck (2009)
Organic acids		
Aconitic acid	*Aspergillus itaconicus*	Dwiarti et al. (2002)
Carlosic acid	*Penicillium charlesii*	Clutterbuck et al. (1935)
Citric acid	*Penicillium citratum, A. niger, Saccharomyces lipolytica*	Dayalan et al. (2011)
Garlic acid	*P. charlesii*	Jacobsen et al. (1978)
Dimethyl pyruvic acid	*A. niger*	Finogenova et al. (2005)
Formic acid	*Rhizopus* sp.	Tanaka et al. (2006)
Gentistic acid	*Penicillium griseofulvum*	Simonart and Wiaux (1960)
Gluconic acid	*P. chyrsogenum*	Moyer and Coghill (1946)
Glycolic acid	*A. niger*	Martin and Steel (1955)
Lactic acid	*Rhizopus* sp.	Tanaka et al. (2006)
Malic acid	*A. niger, A. fumaricus*	Cochrane (1948)
Malonic acid	*A. funiculosum*	Cochrane (1948)
Oxalic acid	*A. niger*	Strasser et al. (1994)
Succinic acid	*Rhizopus* sp.	Song and Lee (2006)

The application of genetic engineering in the development of fungal strain improvement requires fundamental understanding of the metabolism and biochemistry of the strain of interest (Hwang et al., 2003). Through knowledge about specific metabolic pathways, the regulation of metabolism, or structural and functional relationships of critical genes involved in metabolism is essential for the design of genetic improvement strategies; as it provides the rationale for selection of desirable genes and assures that once inserted into a new host, the genes will be appropriately expressed and regulated as predicted. This whole process requires a vital carrier component called vector (Ness et al., 1999). A vector can be defined as a vehicle for transferring DNA from one strain to another and are prepared from bacterial plasmids or sometimes artificially constructed plasmids are used. Plasmids are generally used for transferring DNA because they are small self replicating circular DNA that are stable and relatively easy to isolate, characterize and manipulate in the laboratory. Native plasmids do not naturally possess all the desirable features of a vector (e.g., multiple cloning sites, selectable markers, ability to replicate in several hosts). Therefore, genetic engineering is frequently used to construct multifunctional cloning vectors (Sahm et al., 2000). Once genes have been identified and cloned into the appropriate vector they must be introduced into a viable host. Since, the recombinant DNA is a naked DNA molecule, gene transfer systems based on protoplast transformation and electroporation methods are most applicable in genetic engineering experiments. High transformation efficiencies greatly facilitate screening and identification of appropriate transformants. Electroporation is the preferred transformation procedure for most fungal strains (Andersen et al., 2002).

Recombinant DNA technique is successful when the host cell containing the recombinant genes expresses the protein from the recombinant genes. A significant amount of recombinant protein will not be produced by the host unless expression factors are added. Protein expression depends upon the gene being surrounded by a collection of signals, which provide instructions for the transcription and translation of the gene by the cell. These signals include the promoter, the ribosome-binding site, and the terminator (Jarvis, 2008). Expression vectors, in which the foreign DNA is inserted, contain these signals and are species specific. Production of recombinant proteins in eukaryotic systems generally takes place in yeast and filamentous fungi, but, not in *E. coli*, because, it is unlikely to understand the signals of eukaryotic promoters and terminators and gene containing introns or signals which act as terminators to a bacterial host. This, results in the premature termination and the recombinant protein produced may not be processed as desired, like improper folding or degradation of protein (Ohnishi et al., 2002).

Certain microorganisms are capable to take up naked DNA present in the surrounding medium by a process called transformation. This process involves selection of a piece of DNA from the fungal strain which codes for industrially important product by using restriction enzyme and the same enzyme used to cut the vector and ligated with ligase. Constructed vector contains selectable marker, which allows for the identification of recombined vector within the host DNA. An antibiotic marker is used, so a host cell without the vector dies when exposed to a certain antibiotic; and the host containing the vector will survive because of the presence of the antibiotic resistance gene in the vector (Lee et al., 2000). The vector is introduced into the host cell by the process of transformation. The host cells must be specially prepared to take up the foreign DNA. Selectable markers like antibiotic resistance, color changes, or any other characteristic which can differentiate transformed hosts from untransformed hosts can be identified. Different vectors have different properties to make them suitable to different applications like symmetrical cloning sites, size and high copy number (Yoneda, 1980). In non-bacterial transformation process, microinjection technique is used. The DNA is injected directly into the nucleus of the cell to be transformed. In biolistics, the host cells are bombarded with high velocity micro-projectiles, such as gold particles or tungsten, coated with the recombinant DNA to be integrated into the host genome (Stephanopoulos, 1999).

Saccharomyces cerevisiae is widely used as a eukaryotic model for recombinant DNA technology. One of the main reasons is that, the transmission genetics of yeast is extremely well understood and another important advantage is the availability of a circular 6.3-kb natural yeast plasmid. This plasmid has a circumference of 2 μm and forms the basis for several sophisticated cloning vectors; moreover, this plasmid is transmitted to the cellular products of meiosis and mitosis (Choi et al., 2003). The simplest yeast vectors are derivatives of bacterial plasmids into which the yeast locus of interest has been inserted. When these vectors are transformed into yeast cells, these plasmids get integrated into yeast chromosomes generally by homologous recombination through single or double crossover. As a result, either the entire plasmid is inserted into the genome or the targeted allele is replaced by the allele on the plasmid. Such integrations can be detected by plating cells on a medium that selects for the allele on the plasmid. Because bacterial plasmids do not replicate in yeast, integration is the only way to generate a stable modified genotype (Morrow, 2009). Plasmids can be used to study the regulatory elements present upstream of a gene. The relevant coding region and its upstream region can be spliced and inserted into a plasmid, which can be selected by separate yeast markers (Werten et al., 1999). The upstream region can be manipulated by inducing a series of deletions, which are achieved by cutting the DNA, using special exonucleases to chew away the DNA in one direction to different extents and then rejoining it. The experimental objective is to determine which of these deletions still permits normal functioning of the gene. Proper function is assayed by transforming the plasmid into a recipient in which the chromosome locus carries a defective mutant allele and then monitoring for the return of gene function in the recipient. The results generally determine the specific region that is necessary for gene regulation and the normal functioning of the gene (Mayer et al., 1980).

Fungal strain improvement strategies are used to increase production of lactic acid, antibiotic production, amino acid and enzymes (Table 5.3). In the presence of glucose as substrate, the fungi will produce 60–80% of lactic acid and remaining as ethanol. By using rDNA technology lactic acid production is increased in fungi by transforming the gene *ldhA*, which codes for dehydrogenase, which acts on pyruvate and increases the levels of lactic acid. Recombinant *Saccharomyces cerevisiae* containing six copies of bovine L-lactate dehydrogenase produces 122 gl^{-1} of lactic acid from cane sugar, with optical purity of 99.9%. *Saccharomyces cerevisiae* normally produces 2 gl^{-1} of malic acid from fumaric acid; a recombinant strain containing a cloned fumarase gene was able to produce 125 gl^{-1} with a yield of approximately 90% (Demain and Vaishnav, 2009). With gene shuffling in yeast cells ethanol production is increased, as it is essentially important in research and laboratories. Metabolic engineering of penicillin-producing strains such as *Pseudomonas chrysogenum, Pseudomonas putida* and *Pseudomonas nidulans* to increase

TABLE 5.3 Industrially Important Enzymes Produced from Fungi

Fungi	Enzyme	Function	Industrial applications	References
Aspergillus oryzae, Aspergillus niger	Amylase	Hydrolyses starch to dextrin and sugar	Textile industry, fruit juice clarification, bread making, paper industry, adhesive preparation, syrup preparation, fermentation, brewery and pharmaceutical industries	Aiyer (2005)
Aspergillus sp.	Pectinase	Hydrolyses pectin	Linen industry, coffee bean fermentation and wine industry	Favela-Torres et al. (2006)
Aspergillus sp.	Protease	Hydrolyses protein into simpler compounds	Textile industry, leather industry, meat tenderizing, bread making, digestive aid preparation, detergent manufacturing and glue industry	Rao et al. (1998)
Basidiomycetes sp.	Glucose isomerase	Converts glucose to fructose	Fruit juice manufacturing	Crueger and Crueger (1990)
Rhizopus arrhizus, Mucor javanicus, Candida rugosa, C. cylindracae, C. lipolytica, Pseudomonas aeruginosa, Aspergillus niger, Penicillium cyclopium	Lipase	Converts fats and oils into glycerol and fatty acids	Dairy industry, textile industry, detergent manufacturing, esterification reaction and soap industry	Macrae and Hammond (1985)

TABLE 5.3 Continued

Fungi	Enzyme	Function	Industrial applications	References
Saccharomyces cerevisiae	Lactase	Converts lactose to glucose and galactose	Dairy industry	O'leary et al. (1977)
Saccharomyces cerevisiae	Invertase	Hydrolyzes sucrose to form glucose	Sweets and candy industry and syrup manufacturing	Ostergaard et al. (2000)
Streptomyces sp.	Keratinase	Converts keratin into simpler compounds	Leather industry	Kornillowicz-Kowalska and Bohacz (2011)
Aspergillus oryzae, Trichoderma harziannum, T. ressei, T. viride	Cellulase	Hydrolysis of cellulose	Pulp and paper industry and biofuel production	Reese (1956)

penicillin production. Some of the recombinant products produced by yeast, *P. pastoris* are tumor necrosis factor, gelatin, intracellular tetanus toxin fragment C and serum albumin. Some products obtained from recombinant DNA technology by using *S. cerevisiae* are insulin, hepatitis B surface antigen, urate oxidase, glucagons, granulocyte macrophage colony stimulating factor (GM-CSF), hirudin, and platelet-derived growth factor (Ness et al., 1999).

KEYWORDS

- **Adaptation**
- **Antibiotic marker**
- **Chemical mutagens**
- **Chimeric DNA**
- **Electroporation**
- **Ethyl methane sulfonate**
- **Eukaryotic promoters**
- **Filamentous fungi**
- **Foreign DNA**
- **Gene cloning**
- **Genetic crossing**
- **Genetic engineering**
- **Genetic recombination**
- **Media optimization**
- **Metabolic engineering**
- **Metabolic pathways**
- **Methyl methane sulfonate**
- **Microinjection**
- **Mutagenesis**
- **Mutant isolation**
- **Parasexuality**
- **Post-transitional modification**
- **Protoplast transformation**

- *Pseudomonas chrysogenum*
- *Pseudomonas nidulans*
- *Pseudomonas putida*
- **Random mutagenesis**
- **Recombinant DNA technology**
- **Recombinant proteins**
- **Replica plating**
- *Saccharomyces cerevisiae*
- **Synthetic DNA**
- **Transformation**

REFERENCES

Adrio, J. L., Demain, A. L. (2010). Recombinant organisms for production of industrial products. *Bioeng Bugs.* 1(2), 116–131.

Andersen, D. C., Krummen, L. (2002). Recombinant protein expression for therapeutic applications. *Curr. Opin. Biotechnol.* 13, 117–123.

Becker, T. C. A., Castro-Prado, M. A. A. D. (2006). Parasexuality in asexual development mutants of *Aspergillus nidulans. Biol. Res.* 39, 297–305.

Bradley, S. G. (1962). Parasexual phenomena in microorganisms. *Annual Review of Microbiology* 16, 35–52.

Burg B. V., Vriend, G., Veltman, O. R., Venema, G., Eijsink, V. G. H. (1998). Engineering an enzyme to resist boiling. *Proc. Natl. Acad. Sci. USA* 95, 2056–2060.

Choi, B. K., Bobrowicz, P., Davidson, R. C., Hamilton, S. R., Kung, D. H., Li, H., Miele, R. G., Nett, J. H., Wildt, S., Gerngross, T. U. (2003). Use of combinatorial genetic libraries to humanize N-linked glycosylation in the yeast *Pichia pastoris. Proc. Natl. Acad. Sci. USA* 100, 5022–5027.

Demain, A. L., Vaishnav, P. (2009). Production of recombinant proteins by microbes and higher organisms. *Biotechnol. Adv.* 27, 297–306.

Dufossé, L., Fouillaud, M., Caro, Y., Mapari, S. A. S., Sutthiwong, N. (2014). Filamentous fungi are large-scale producers of pigments and colorants for the food industry. *Current Opinion in Biotechnology* 26, 56–61.

Eggeling, L., Sahm, H., de Graaf, A. A. (1996). Quantifying and directing metabolic flux: application to amino acid overproduction. *Adv Biochem Eng Biotechnol.* 54, 1–30.

Flickweert, M. T., Kuyper, M., van Maris, A. J. A., Kotter, P., van Dijken, J. P., Pronk, J. T. (1999). Steady-state and transient-state analysis of growth and metabolic production in a *Saccharomyces cerevisiae* strain with reduced pyruvate-decarboxylase activity. *Biotechnology and Bioengineering* 66(1), 42–50.

Hopwood, D. A., Bibb, M. J., Chater, K. F., Kieser, T., Bruton, C. J., Kieser, H. M., Lydiate, D. J., Smith, C. P., Ward, J. M., Schrempf, H. (1985). *Genetic manipulation of Streptomyces: a laboratory manual*, The John Innes Foundation, Norwich, UK.

Hwang, Y. S., Kim, E. S., Biró, S., Choi, C. Y. (2003). Cloning and analysis of a DNA fragment stimulating avermectin production in various *Streptomyces avermitilis* strains. *Appl Environ Microbiol.* 69, 1263–1269.

Jarvis, L. M. (2008). A technology bet. DSM's pharma product unit leverages its biotech strength to survive in a tough environment. *Chem Eng News* 86, 30–31.

Lee, J. Y., Hwang, Y. S., Kim, S. S., Kim, E. S., Choi, C. Y. (2000). Effect of a global regulatory gene, *afsR2*, from *Streptomyces lividans* on avermectin production in *Streptomyces avermitilis*. *Biosci Bioeng.* 89, 606–608.

Mayer, H., Collins, J., Wagner, F. (1980). Cloning of the penicillin G-acylase gene of *Escherichia coli* ATCC 11105 on multicopy plasmids. *In Enzyme Engineering, Vol. 5*, ed. by Weetall, H. H., Royer, G. P., Plenum Press, New York, pp. 61–69.

Meyer, H. P., Biass, J., Jungo, C., Klein, J., Wenger, J., Mommers, R. (2008). An emerging star for therapeutic and catalytic protein production. *BioProc Internat.* 6, 10–21.

Morrow, K. J. (2009). Grappling with biologic manufacturing concerns. *Gen Eng Biotechnol News.* 29, 54–55.

Ness, J. E., Welch, M., Giver, L., Bueno, M., Cherry, J. R., Borchert, T. V., Stemmer, W. P., Minshull, J. (1999). DNA shuffling of subgenomic sequences of subtilisin. *Nat Biotechnol.* 17, 893–896.

Nevalainen, K. M. H., Valentino, S. J. T., Bergquis, P. L. (2005). Heterologous protein expression in filamentous fungi. *Trends Biotechnol.* 23, 468–474.

Nevoigt, E. (2008). Progress in metabolic engineering of *Saccharomyces cerevisiae*. *Microbiol Mol Biol Rev.* 72(3), 379–412.

Nielsen, J. (2001). Metabolic engineering. *Appl Microbiol Biotechnol.* 55, 263–283.

Ohnishi, J., Mitsuhashi, S., Hayashi, M., Ando, S., Yokoi, H., Ochiai, K., Ikeda, M. (2002). A novel methodology employing *Corynebacterium glutamicum* genome information to generate a new L-lysine-producing mutant. *Appl Microbiol Biotechnol.* 58, 217–223.

Parekh, S., Vinci, V. A., Strobel, R. J. (2000). Improvement of microbial strains and fermentation processes. *Appl Microbiol Biotechnol.* 54, 287–301.

Peberdy, J. F. (1985). Mycolytic enzymes. *In Fungal protoplast: applications in biochemistry and genetics*, ed. by Peberdy, J. F., Ferenczy, L., Marcel Dekker, New York, pp. 31–34.

Ryu, D. D. Y., Kim, K. S., Cho, N. Y., Pai, H. S. (1983). Genetic recombination in *Micromonospora rosaria* by protoplast fusion. *Appl Environ Microbiol.* 45, 1854–1858.

Sahm, H., Eggeling, L., de Graaf, A. A. (2000). Pathway analysis and metabolic engineering in Corynebacterium glutamicum. *Biol. Chem.* 381, 899–910.

Schneiter, R. (2004). *Genetics, Molecular and Cell Biology of Yeast*, Universität Freiburg Schweiz, Switzerland.

Sharma, R., Katoch, M., Srivastava, P. S., Qazi, G. N. (2009). Approaches for refining heterologous protein production in filamentous fungi. *World J Microbiol Biotechnol.* 25(12), 2083–2094.

Stephanopoulos, G. (1999). Metabolic fluxes and metabolic engineering. *Metab Eng.* 1, 1–11.

Suenaga, H., Mitsokua, M., Ura, Y., Watanabe, T., Furukawa, K. (2001). Directed evolution of biphenyl dioxygenase: emergence of enhanced degradation capacity for benzene, toluene and alkylbenzenes. *J. Bacteriol.* 183, 5441–5444.

Tolmsoff, W. J. (1983). Heteroploidy as a mechanism of variability among fungi. *Annual Review of Phytopathology* 21, 317–340.

Valent, B., Chumley, F. G. (1991). Molecular genetic analysis of the rice blast fungus, *Magnaporthe grisea*. *Annual Review of Phytopathology* 29, 443–467.

Virgin, J. B., Bailey, J. P., Hasteh, F., Neville, J., Cole, A., Tromp, G. (2001). Crossing over is rarely associated with mitotic intragenic recombination in *Schizosaccharomyces pombe*. *Genetics* 157, 63–77.

Werten, M. W. T., van den Bosch, T. J., Wind, R. D., Mooibroek, H., de Wolf, F. A. (1999). High-yield secretion of recombinant gelatins by *Pichia pastoris*. *Yeast* 15, 1087–1096.

Xu, F., Jin, H., Li, H., Tao, L., Wang, J., Lv, J., Chen, S. (2011). Genome shuffling of *Trichoderma viride* for enhanced cellulose production. *Ann. Microbiol.* 5, 176–191.

Yoneda, Y. (1980). Increased production of extracellular enzymes by the synergistic effect of genes introduced into *Bacillus subtilis* by stepwise transformation. *Appl Environ Microbiol* 39, 274–276.

CHAPTER 6

MOLECULAR IMPROVEMENT OF FUNGAL STRAINS

CONTENTS

6.1   INTRODUCTION

Genetics plays a significant role in the strain improvement of industrially important microorganisms, whether the species is prokaryotic or eukaryotic. To carry out any of the molecular genetic method for strain improvement, identification of biosynthetic pathway, adequate vectors and appropriate transformation protocols should be developed. Among microorganisms, fungi and actinomycetes have been successfully developed for industrial applications. Several species of filamentous fungi are known to produce pharmaceuticals, antibiotics, metabolites, phytohormones, and other industrially important products. There is a great deal of interest to utilize the potential of filamentous fungi as biocontrol agents, antagonists of other fungal phytopathogens, bioherbicides and as bioinsecticides (Kapoor, 1995). Genetic engineering methods have also provided tools to know in detail the nature of the modifications occurred (Barrios-González, 2003). Genetic technology can improve the performance of microorganisms by altering spectrum of metabolites as well as altering the species growth characteristics. In addition, microbes can be made to produce novel products by rDNA technology with appropriate screening. Genetic manipulations are employed in industries to improve the production rate and quality by the originally isolated strain. Today, classical method has been replaced by modern strategic technologies developed via advances in molecular biology, recombinant DNA technology, and genetics to improve strain. The implementation of strain improvement methods has increased fermentation productivity and decreased costs tremendously. Additionally, these genetic programs also serve other goals such as the elimination of undesirable products or analogs, discovery of new antibiotics, and deciphering of biosynthetic pathways (Demain and Adrio, 2008). Previously, scientists selected microbial strains from natural ecosystem that fulfilled both microbiological and technical requirements for economical production processes. Later, genetically modified strains with novel properties were developed through classical strain improvement methods relying mostly on random mutagenesis. In recent years, recombinant technologies have contributed extensively to production enhancement. The design and generation of new transformation strategies to manipulate specific gene expression and function in diverse filamentous fungi, including those having a biotechnical significance has been achieved by better understanding of fundamental genetic processes. With developments in genetic engineering and molecular biology, filamentous fungi have also achieved increased attention as hosts for recombinant DNA. Considerable success has been achieved in the increased production of variety of fungal products such as secondary metabolites, enzymes and proteins (Schwab, 1988; Archer and Peberty, 1997; Wang et al., 2005; Kuck and Hoff, 2010).

Knowledge on fungal genetics and their biochemical pathways has been important for application in strain improvement. Normally, filamentous fungi are transformed to get potentially superior long-term stability of the fungal transformants by plasmids that integrate into the fungal genome. For these reasons, filamentous fungi have tremendous potential as hosts of recombinant DNA. The transformed strains of species such as *Aspergillus* sp., *Trichoderma reesei*, *Chrysosporium lucknowense* and *Mortierella alpinis* have been used as a host for the production of recombinant proteins (Harkki et al., 1989; Gouka et al., 1997; Hombergh et al., 1997; Mach and Zeilinger, 2003). Genetic manipulation has also successfully been employed to improve the biocontrol ability of biocontrol agents. The choice of a commercial fungal strain, which produce high amounts of target proteins and metabolites by high-level expression of the desired genes, depends on growth conditions, the level of desired gene expression, post-translational modifications and biological activity of the desired metabolites (Archer and Peberdy, 1997).

Filamentous fungi have become indispensable for the production of enzymes of fungal and non-fungal origin and also have exceptionally high capacity to express and secrete proteins. Currently, *Aspergillus niger*, *Aspergillus oryzae* and *Trichoderma reesei* are developed for the production of native and recombinant enzymes (Punt et al., 2002). In addition, filamentous fungi are producing many secondary metabolites in which few are biotechnologically produced and utilized mainly in pharmaceutical industries. The β-lactam group of antibiotics, including penicillin and cephalosporin, was the first group that benefited from the progress made in molecular techniques for filamentous fungi (Brakhage and Caruso, 2004). Some other bioactive compounds produced by filamentous fungi and playing important role for human welfare are cyclosporine A (immunosuppressive agent), lovastatin (cholesterol-lowering agent), taxol (anticancer agent) and griseofulvin (antifungal agent) (Archer, 2000; Willke and Vorlop, 2001; Polizeli et al., 2005; Olempska-Beer et al., 2006; Meyer, 2008). Adequate knowledge and experience to perform techniques using molecular tools are important (Davati and Najafi, 2013).

6.2 GENETIC RECOMBINATION THROUGH TRANSFORMATION

Genetic recombination can also be done by transformation, which involves the uptake of part of DNA from donor organism and to insert into the recipient organism (Das et al., 2011). Most filamentous fungi are transformed by plasmids that integrate into the fungal genome, suggesting potentially superior long-term stability of the fungal transformants. For these reasons, filamentous fungi have tremendous potential as hosts of recombinant DNA (Wiebe, 2003). Several efficient transformation methods are being developed for large number of fungal species. In filamentous fungi, protein secretion plays an important role and acts as host for rDNA. The genus *Aspergillus* has been successfully used as a host for the production of recombinant proteins such as glucoamylase, bovine chymosin, human lactoferrin, hen egg-white lysozyme, human interleukin-6, and thaumatin. Other than *Aspergillus*, fungal species such as *Trichoderma reesei* and *Chrysosporium lucknowense* are also been used as a host strain (Gouka et al., 1997; Hombergh et al., 1997; Maras et al., 1999; Mach and Zeilinger, 2003; Meyer, 2008). Foremost, the development of transformation method is a major part for many fungal species.

Transformation technique is also important in the selection of mode of frequency of individual integration events from homologous or illegitimate recombination. Thus, designing of a genetic engineering strategy primarily requires consideration of the most suitable transformation method (Meyer, 2008). In the earlier stage, the use of protoplasts for transformation has been extended to several filamentous fungi. However, the frequency of transformation is extremely low when compared to yields obtained with *Saccharomyces cerevisiae*. In order to improve transformation of filamentous fungi, progress has been made with the establishment of alternative methods for fungal transformation such as electroporation, biolistic transformation and *Agrobacterium* mediated transformation (AMT) (Ruiz-Diez, 2002; Michielse et al., 2005). These methods have especially been valuable for fungal strains that do not form sufficient numbers of protoplasts or whose protoplasts do not regenerate sufficiently (MacKenzie et al., 2004).

6.3   PROTOPLAST FUSION IN FUNGI

Protoplast fusion is a new versatile technique to endorse genetic recombination in a variety of prokaryotic and eukaryotic cells in specific with the production of interspecific or intergeneric hybrids (Bhojwani et al., 1977; Murlidhar and Panda, 2000). Additionally, low frequency of recombination created a way to the use of protoplast fusion in strain improvement techniques. Since 1981, the application of genetic recombination in the production of important microbial products is increased. Today, strain improvement is routinely been achieved by protoplast fusion between different mutant lines (Ryu et al., 1983). It has become an important tool of gene manipulation as it breaks down the barriers to genetic exchange by conventional methods.

Degradation of cell wall of fungi is important to carryout protoplast fusion. Commonly, fungal wall is degraded by Novozyme-234, which includes glucanase and chitinase that are added rapidly to the growing fungi suspended in an osmotic buffer (Narayanswamy, 1994; Jogdand, 2001). The two important criteria to be considered for proper generation of protoplasts are the age of mycelia and contact time with the lytic enzyme. The protoplast and DNA are mixed in 15% (w/v) PEG 6000 which causes clump formation and pH buffer (Tris-HCl). Grown mycelium is inoculated on agar and incubated at room temperature for overnight. After overnight incubation, enzyme will be added and incubated at 30°C for 15 hours in a petridish containing KCl, filtered and then the protoplast is washed with KCl, followed by centrifugation and then the pellets are resuspended. Normally, frequency of protoplast fusion in fungi is 0.2–2.0% (Srinivas and Panda, 1997; Jogdand, 2001). Currently, most of the laboratories are involved in fungal gene manipulation using protoplast fusion.

The alternative approaches for the direct bioconversion of ethanol from cellulosic material by the intergeneric fusants between *Trichoderma reesei* and *Saccharomyces cerevisiae* appears to be the best technique. Also, the production of complete set of cellulase is achieved by the protoplast fusion of *Trichoderma reesei*, which is responsible for the production of endo and exoglucanase and *Aspergillus niger* which produces β-glucosidase (Ahmed and Berkley, 2006). Prabhavathy et al. (2006) reported that the isolated protoplast from *Trichoderma reesei* strain PTr2 showed high CMCase activity with 80% of fusants and more than two fold increments in enzyme activities with two fusants SFTr2 and SFTr3 as compared to the parental strain PTr2. Intraspecific protoplast fusion was carried out by Nazari et al. (2005) in *Streptomyces griseoflavus* with PEG 1000 to increase the production of desferrioxamine B chelator which is used in the medicine to absorb additional iron from the blood of thalasemia patients. Also, this technique is applied between *Saccharomyces diastatiacus* and *Saccharomyces uvarum* strain to improve the flavor of beer (Janderova et al., 1990). Protoplast fusion between *Helminthosporium gramineum* and *Curvularia lunata* was carried out by Zhang et al. (2007) to breed new strains with improved spore productivity and obtained fusant strains had increased production of the phytotoxin, ophiobolin A. Protoplast fusion of morphologically and biochemically different *Claviceps purpurea* strains producing exotoxins was carried out by Brume et al. (1992) which produced tenfold higher alkaloid than parental strains. Fusants of *Penicillium chrysogenum* and *Cephalosporium acremonium* produced a novel lactam antibiotic. Normally, *Penicillium chrysogenum* shows less sporulation and poor seed growth which is improved by protoplast fusion to produce more antibiotic in shorter period and better growth in seed medium (Lein, 1986; Skatrud et al., 1989).

Stable hybrids fermenting high concentration of glucose (49% w/w) was produced by interspecific protoplast fusion between osmotolerant *Saccharomyces mellis* and the highly fermentative *Saccharomyces cerevisiae* (Legmann and Margalith, 1983; Adrio and Demain, 2006). Another application of protoplast fusion is the recombination of improved producers from a single mutagenesis treatment. Two different strains of *Nocardia* producing cephamycin C were fused by Wesseling and Lago (1981) to increase the yield of 10–15% more antibiotic than the best parent. This technique allows the discovery of new antibiotics by fusing producers of different or even the same antibiotics. A recombinant obtained from two different rifamycin-producing strains of *Nocardia mediterranei* produced two new rifamycins (16,17-dihydrorifamycin S and 16,17-dihydro-17-hydroxy-rifamycin S) (Traxler et al., 1982). Okanishi et al. (1996) carried out the interspecific protoplast fusion to yield recombinants between *Streptomyces griseus* and five other species such as *Streptomyces cyaneus*, *Streptomyces exfoliatus*, *Streptomyces griseoruber*, *Streptomyces purpureus* and *Streptomyces rochei*, of which, 60% of fusants produced no antibiotics and 24% produced antibiotics different from the parent strains. Changing the order of an individual pathway in a parent is used to produce new antibiotics (Hershberger, 1996; Adrio and Demain, 2006). Protoplast fusion between non-antibiotic producing mutants of *Streptomyces griseus* and

Streptomyces tenjimariensis has generated a new antibiotic indolizomycin (Gomi et al., 1984). Osmotolerance of food yeasts, such as *Saccharomyces cerevisiae* and *Saccharomyces diastaticus*, was increased by protoplast fusion with osmotolerant yeasts. Other traits transferred between yeasts by protoplast fusion include flocculation (Panchal et al., 1982), lactose utilization (Farahnak et al., 1986), the killer character (Bortol et al., 1986; Farris et al., 1992), cellobiose fermentation (Pina et al., 1986) and methionine overproduction (Brigidi et al., 1988; Adrio and Demain, 2006).

6.4 GENETIC TRANSFORMATION BY ELECTROPORATION

Electroporation is a rapid and simple technique for introducing cloned genes into a wide variety of organisms for the generation of transient expression. Targeted cells are subjected to electric shock; nanometer-sized pores in the membrane will be created due to electric shock, during which exogenous DNA can enter into the cell from the suspending solution. Basically, electroporation is based on the reversible permeabilization of biomembranes induced by high amplitude electric fields for short duration. During an electric pulse, membrane permeability changes, which permit the uptake of recombinant DNA, in turn, can result in molecular transformation (Richey, 1989; Kapoor, 1995; Weaver, 1995; Prabha and Punekar, 2004). A proportion of these cells become stably transformed and can be selected if a suitable marker gene is carried on the transforming DNA. Temperature, voltage, resistance and capacitance of electric field, topological form of the DNA, host genetic background and growth condition are the important factors which determine the efficacy of the electroporation (Hanahan et al., 1991; Primrose and Twyman, 2006). This method has been used as an alternative means of DNA transformation for animal cells, plant protoplasts, yeast, bacteria and filamentous fungi (Chakraborty et al., 1991; Ozeki et al., 1994; Riach and Kinghorn, 1996; Ruiz-Diez, 2002). In filamentous fungi, electroporation technique has been developed for transformation of *Neurospora crassa*, *Penicillium urticae*, *Leptosphaeria maculans*, *Aspergillus oryzae* (Chakraborty et al., 1991), *Aspergillus nidulans* (Sanchez and Aguirre, 1996), *Aspergillus niger* (Ozeki et al., 1994), *Wangiella dermatitidis* and *Aspergillus fumigatus* (Kwon-Chung et al., 1998).

6.5   BIOLISTIC TRANSFORMATION

Efficient uptake of foreign DNA using transformation techniques is not applicable to all filamentous fungi. In some of the fungi, production of protoplast has been a critical method. To overcome this issue, biolistic method was introduced by Klein et al. (1987). It has been developed as a method for incorporation of plasmid DNA into intact, thick-walled fungal cells. Gold coated DNA or tungsten particles at high velocity are aimed directly to get into fungal spores or hyphae. The DNA dissociates from the coated particles and moves towards the nucleus and integrated with the genome (Ruiz-Diez, 2002; Prabha and Punekar, 2004). Biolistics method creates transformation in numerous fungi such as *Neurospora crassa, Magnaporthe grisea, Trichoderma harzianum, Aspergillus nidulans, Gliocladium virens, Paxillus involutus,* and *Trichoderma reesei* (Lorito et al., 1993; Bills et al., 1995; Riach and Kinghorn, 1996; Gomes-Barcellos et al., 1998; Hazell et al., 2000).

6.6   AGROBACTERIUM MEDIATED TRANSFORMATION

Agrobacterium mediated transformation (AMT) system can also be used for the transformation of several filamentous fungi, such as *Agaricus bisporus, Calonectria morganii, Fusarium circinatum* and *Helminthosporium turcicum,* which were recalcitrant to other transformation methods (Groot et al., 1998). It is an efficient tool for transformation of filamentous fungi (Groot et al., 1998; Amey et al., 2002; Fitzgerald et al., 2003; Meyer et al., 2003). The ability of *Agrobacterium tumefaciens* to transfer its DNA to fungi belonging to various classes is indicative of the potential of this transformation system for fungal biotechnology. AMT has several advantages over conventional transformation like intact cells such as conidia, vegetative and fruiting body mycelia can be used as starting material; thereby, eliminating the need of protoplast generation which is laborious.

Various strains of *Agrobacterium* such as LBA4404, EHA105 and LBA1100 have been used for the transformation of fungi and oomycetes. However, the usage of *Agrobacterium* strains derived from the supervirulent A281 strain which shows high level of *vir* gene expression resulted in higher transformation frequencies in *Saccharomyces cerevisiae, Monascus purpureus*, and Oomycete *Phytophthora infestans*, compared with AMT using *Agrobacterium* strain LBA1100 (Piers et al., 1996; Campoy et al., 2003; Vijn and Govers, 2003). Transformation of T-DNA to *Cryphonectria parasitica* was more efficient by supervirulent *Agrobacterium* A281 strain and its derivative AGL-1 than the *Agrobacterium* LBA4404 strain (Park and Kim, 2004). Although, it is not possible to point out which *Agrobacterium* strains are most suitable for AMT of fungi, it is clear that the choice of *Agrobacterium* strain can definitely effect on transformation frequency (Michielse, 2005).

6.7 SINGLE CHROMOSOME TRANSFER

Several species of filamentous fungi contains supernumerary chromosomes. These chromosomes are dispensable for the fungus to survive, but, may carry genes required for specialized functions. Some dispensable chromosomes are able to transfer horizontally or in the absence of a sexual cycle from one fungal strain to another. Does and Rep (2012) described the horizontal chromosome transfer (HCT) occurs during incubation of two strains in a same plate. Selection of HCT progeny is necessary for the actual occurrence of HCT. Two different fungal strains which are having genes important for industrial production can be incubated as co-incubation. HCT can be initiated with equal amounts of asexual spores of both strains, plated on regular growth medium for the particular fungus, followed by incubation until new asexual spores are formed. The new asexual spores are then harvested and plated on plates containing the target compound (may be for degradation or utilization). Capable or resistant colonies that appear should carry at least one chromosome from each parental strain. Finally, strains carrying two characteristics need to be analyzed to assess whether HCT has actually occurred or not. This can be done by various genome-mapping methods, like CHEF-gels, AFLP, RFLP, PCR markers, optical maps, or even complete genome sequencing.

6.8 GENOME SHUFFLING

Genome shuffling is an emerging and highly effective method, which is widely used in microbiology for the production of metabolites, improving substrate uptake as well as enhancing strain tolerance. The genetic breeding can be performed on the target microorganisms without knowing its genetic background and multiparental crossing is allowed by shuffling DNA together with the recombination of entire genomes normally associated with conventional genome that increases recombination process (Powell et al., 2001; Stemmer, 2002; Cheng et al., 2009; Gong et al., 2009; Leja et al., 2011). This technique was first employed by Zhang et al. (2002) to improve the ability of *Streptomyces fradiae* to synthesize Tailexing. *Nodulisporium sylviform* was used as starting strain to apply the fundamental principles of genome shuffling in breeding of taxol producing fungi by Zhao et al. (2008). Three hereditarily stable strains with high taxol production were obtained by four cycles of genome shuffling. In conclusion, a high taxol producing fungus, *Nodulisporium sylviform* F4–26, was obtained, which produced 516.37 µg/L taxol. This value is 64.41% higher than that of the starting strain NCEU-1 and 31.52%-44.72% higher than that of the parent strains (Zhao et al., 2008). This technique is similar to the classical strain improvement, the main difference between these two techniques is that, genome shuffling process is sexual, in which populations improved strains are evolved than the classical method and as a result, the final improved strains involve the genetic trait from multiple initial strains (Cardayré and Powell, 2003). Through this technique, different genes which are associated with the production of metabolites can be recombined during several rounds of genome shuffling and consequently desirable phenotypes can be obtained (Jin et al., 2009). Nowadays, the technique of genome shuffling is used to significantly improve the quality of industrially important microbiological phenotypes. It is quite cost effective method which does not require any expensive facility and can be easily employed in most of the laboratories (Gong et al., 2009). In addition, shuffled strains are not considered as genetically modified organisms and hence, can be used in the food industry (Zhang et al., 2002; Ahmed, 2003; Petri and Dannert, 2004). Genome shuffling can be integrated with metabolic engineering to facilitate the creation of complex phenotypes, thereby, increasing the metabolite yield (Patnaik et al., 2002; Dai and Copley, 2004; Bode and Muller, 2006; Liang and Guo, 2007; Otte et al., 2009; Zhang et al., 2010), through improved substrate uptake and conversion (Kumar, 2007; John et al., 2008), and enhance strain tolerance (Shi et al., 2009; Cui et al., 2014). Lipase production of *Penicillium expansum* was improved by genome shuffling between a lipase-producing mutant strain *Penicillium expansum* FS8486 and a wild type *Aspergillus tamari* FS-132 isolated from the soil of a volcano in Xinjiang in China were used as parental strains. After two rounds of genome shuffling, daughter strains with desirable features were screened. Lipase activity in one of the daughter strains was increased over the starting strain FS8486 (Lin et al., 2007). Gong et al. (2007) mutated the epothilone that produced myxobacterium *Saccharomyces cellulosum* strain So0157–2 to improve the production of epothilones, highly promising prospective anticancer agents. The epothilone production of fusants increased about 130 times after two rounds of genome shuffling as compared to the starting strain. In combination with an appropriately designed screening strategy, the mutant *Pichia stipitis* TJ2–3 strain was obtained in a study carried out by Shi et al. (2014). This yeast mutant had enhanced rates of xylose consumption and ethanol production, as compared with the parental strains.

Improvement of cellulase production of *Penicillium decumbens* by genome shuffling of an industrial catabolite-repression-resistant strain JU-A10 with its mutants was carried out by Cheng et al. (2009). The fusants obtained after genome shuffling could produce abundant cellulase much earlier, and they could be potential candidates for bioconversion process. Thus, genome shuffling is an effective and convenient tool in recombinant technology that combines the advantage of multiparental crossing allowed by DNA shuffling with the recombination of entire genomes normally associated with conventional breeding. It is a better technology for genetic engineering and metabolic engineering at the whole genome level. The application of this method does not require expensive facilities. The cost of genome shuffling is not high, either. Importantly, this technique is easy to handle and can be used in most laboratories.

KEYWORDS

- Actinomycetes
- *Agaricus bisporus*
- *Agrobacterium* mediated transformation
- *Aspergillus nidulans*
- Biolistic transformation
- *Calonectria morganii*
- Cellobiose fermentation
- *Chrysosporium lucknowense*
- Co-incubation
- Cyclosporine A
- DNA shuffling
- Electroporation
- Epothilones
- Ethanol production
- Fungal biotechnology
- Fungal genome
- Fungal transformants
- *Fusarium circinatum*
- Genetic breeding
- Genetic recombination
- Genetically modified organisms
- Genome sequencing
- Genome shuffling
- *Gliocladium virens*
- Glucoamylase
- Griseofulvin
- *Helminthosporium turcicum*
- Horizontal chromosome transfer
- Lovastatin
- *Magnaporthe grisea*

- *Mortierella alpinis*
- *Neurospora crassa*
- *Nocardia mediterranei*
- *Nodulisporium sylviform*
- *Paxillus involutus*
- *Penicillium chrysogenum*
- *Penicillium decumbens*
- *Penicillium expansum*
- **Phytohormones**
- *Pichia stipitis*
- **Protoplast fusion**
- **Random mutagenesis**
- **Recombinant proteins**
- *Saccharomyces cellulosum*
- *Saccharomyces diastatiacus*
- *Saccharomyces uvarum*
- **Single chromosome transfer**
- *Streptomyces fradiae*
- *Streptomyces griseoflavus*
- **Taxol**
- *Trichoderma harzianum*
- *Trichoderma reesei*
- **Xylose consumption**

REFERENCES

Adrio, J. L., Demain, A. L. (2006). Genetic improvement of processes yielding microbial products. *FEMS Microbiol. Rev.* 30, 187–214.

Ahmed, F. E. (2003). Genetically modified probiotics in foods. *Trends Biotechnol.* 21, 491–497.

Amey, R. C., Athey-Pollard, A., Burns, C., Mills, P. R., Bailey, A., Foster, G. D. (2002). PEG-mediated and *Agrobacterium*-mediated transformation in the mycopathogen *Verticillium fungicola*. *Mycol. Res.* 106, 4–11.

Archer, D. B. (2000). Filamentous fungi as microbial cell factories for food use. *Curr. Opin. Biotechnol.* 11, 478–483.

Archer, D. B., Peberty, J. F. (1997). The molecular biology of secreted enzyme production by fungi. *Crit. Rev. Biotechnol.* 17, 273–306.

Barrios-González, J., Fernández, F. J., Tomasini, A. (2003). Microbial secondary metabolites production and strain improvement. *Indian Journal of Biotechnology* 2, 322–333.

Bhojwani, S. S., Powar, J. B., Cocking, E. L. (1977). Isolation, culture and division of protoplast. *Plant Sci. Lett.* 8, 85–89.

Bills, S. N., Raicher, D. L., Podila, G. K. (1995). Genetic transformation of the ectomycorrizal fungus *Paxillus involutus* by particle bombardment. *Mycol. Res.* 99, 557–561.

Bode, H. B., Muller, R. (2006). Analysis of myxobacterial secondary metabolism goes molecular. *J. Ind. Microbiol. Biotechnol.* 33, 577–588.

Bortol, A., Nudel, C., Fraile, E., de Torres, R., Giulietti, A., Spencer, J. F. T., Spencer, D. (1986). Isolation of yeast with killer activity and its breeding with an industrial baking strain by protoplast fusion. *Appl Microbiol Biotechnol.* 24, 414–416.

Brakhage, A. A., Caruso, M. L. (2004). Biotechnical genetics of antibiotic biosynthesis. *In The Mycota II. Genetics and Biotechnology*, ed. by Esser, K., Springer-Verlag, New York, pp. 317–353.

Brigidi, P., Matteuzzi, D., Fava, F. (1988). Use of protoplast fusion to introduce methionine overproduction into *Saccharomyces cerevisiae*. *Appl. Microbiol. Biotechnol.* 28, 268–271.

Campoy, S., Perez, F., Martin, J. F., Gutierrez, S., Liras, P. (2003). Stable transformants of the azaphilone pigment-producing *Monascus purpureus* obtained by protoplast transformation and *Agrobacterium*-mediated DNA transfer. *Curr Genet* 43, 447–452.

Cardayré, S. B., Powell, K. (2003). DNA shuffling for whole cell engineering. *In Handbook of Industrial Cell Culture: Mammalian, Microbial, and Plant Cells*, ed. by Vinci, V. A., Parekh, S. R., Humana Press, New Jersey, pp. 476–480.

Chakraborty, B. N., Patterson, N. A., Kapoor, M. (1991). An electroporation-based system for high-efficiency transformation of germinated conidia of filamentous fungi. *Can. J. Microbiol.* 37, 858–863.

Cheng, Y., Song, X., Qin, Y., Qu, Y. (2009). Genome shuffling improves production of cellulose by *Penicillium decumbens* JU-A10. *J. Appl. Microbiol.* 107, 1837–1846.

Cui, Y. X., Liu, J. J., Liu, Y., Cheng, Q. Y., Yu, Q., Chen, X., Ren, X. D. (2014). Protoplast fusion enhances lignocellulolytic enzyme activities in *Trichoderma reesei*. *Biotechnol Lett.* 36(12),2495–2499.

Dai, M. H., Copley, S. D. (2004). Improved production of ethanol by novel genome shuffling in *Saccharomyces cerevisiae*. *Appl. Environ. Microbiol.* 70, 2391–2397.

Das, S., Singh, S., Sharma, V., Soni, M. L. (2011). Biotechnological applications of industrially important amylase enzyme. *International Journal of Pharma and Biosciences* 2(1), 486–496.

Davati, N., Najafi, M. B. H. (2013). Overproduction strategies for microbial secondary metabolites: a review. *International Journal of Life Science and Pharma Research* 3(1), 23–37.

Demain, A. L., Adrio, J. L. (2008). Strain improvement for production of pharmaceuticals and other microbial metabolites by fermentation. *Prog. Drug. Res.* 65, 253–289.

Does H. C. V., Rep, M. (2012). Horizontal transfer of supernumerary chromosomes in fungi. *Methods Mol. Biol.* 835, 427–437.

Farahnak, F., Seki, T., Ryu, D. D. Y., Ogrydziak, D. (1986). Construction of lactose-assimilating and high-ethanol-producing yeasts by protoplast fusion. *Appl Environ Microbiol.* 51, 362–367.

Farris, G. A., Fatichenti, F., Bifulco, L., Berardi, E., Deiana, P., Satta, T. (1992). A genetically improved wine yeast. *Biotechnol Lett* 14, 219–222.

Fitzgerald, A. M., Mudge, A. M., Gleave, A. P., Plummer, K. M. (2003). *Agrobacterium* and PEG-mediated transformation of the phytopathogen *Venturia inaequalis*. *Mycol Res.* 107, 803–810.

Gomes-Barcellos, F., Pelegrinelli-Fungaro, M. H., Furlaneto, M. C., Lejeune, B., Pizzirani-Kleiner, A. A., Azevedo, J. L. (1998). Genetic analysis of *Aspergillus nidulans* unstable transformants obtained by the biolistic process. *Can. J. Microbiol.* 44, 1137–1141.

Gomi, S., Ikeda, D., Nakamura, H., Naganawa, H., Yamashita, F., Hotta, K., Kondo, S., Okami, Y., Umezawa, H., Iitaka, Y. (1984). Isolation and structure of a new antibiotic, indolizomycin, produced by a strain SK2–52 obtained by interspecies fusion treatment. *J Antibiot.* 37, 1491–1494.

Gong, G. L., Sun, X., Liu, X. L., Hu, W., Cao, W. R., Liu, H., Liu, W. F., Li, Y. Z. (2007). Mutation and a high-throughput screening method for improving the production of Epothilones of Sorangium. *J. Ind. Microbiol. Biotechnol.* 34, 615–623.

Gong, J., Zheng, H., Wu, Z., Chen, T., Zhao, X. (2009). Genome shuffling: progress and applications for phenotype improvement. *Biot. Adv.* 27, 996–1005.

Gouka, R. J., Punt, P. J., van den Hondel, C. A. M. J. J. (1997). Efficient production of secreted proteins by *Aspergillus*: progress, limitations and prospects. *Appl Microbiol Biotechnol.* 47, 1–11.

Groot, D., Bundock, P., Hooykaas, P. J., Beijersbergen, A. G. (1998). *Agrobacterium tumefaciens*-mediated transformation of filamentous fungi. *Nat Biotechnol* 16, 839–842.

Hanahan, D., Jessee, J., Bloom, F. R. (1991). Plasmid transformation of *Escherichia coli* and other bacteria. *Methods Enzymol.* 204, 63–113.

Harkki, A., Uusitalo, J., Bailey, M., Penttilä, M., Knowles, J. K. C. (1989). A novel fungal expression system: secretion of active calf chymosin from the filamentous fungus *Trichoderma reesei*. *Bio/Technology* 7, 596–603.

Hazell, B. W., Teo, V. S., Bradner, J. R., Bergquist, P. L., Nevalainen, K. M. (2000). Rapid transformation of high cellulose producing mutant strains of *Trichoderma reesei* by microprojectile bombardment. *Letters in Applied Microbiology* 30, 282–286.

Hershberger, C. L. (1996). Metabolic engineering of polyketide biosynthesis. *Curr Opin Biotechnol.* 7, 560–562.

Hombergh, J. P. T. W. V., van den Vondervoort, P. J. I., Fraissinet-Tachet, L., Visser, J. (1997). *Aspergillus* as a host for heterologous protein production: the problem of proteases. *Trends Biotechnol.* 15, 256–263.

Janderová, B., Cvrčková, F., Bendová, O. (1990). Construction of the dextrin-degrading *pof* brewing yeast by protoplast fusion. *J Basic Microbiol.* 30(7), 499–505.

Jin, Z. H., Xu, B., Lin, S. Z., Jin, Q. C., Cen, P. L. (2009). Enhanced production of spinosad in *Saccharomyces spinosa* by genome shuffling. *Appl. Biochem. Biotechnol.* 159, 655–663.

Jogdand, S. N. (2001). *Protoplast technology, gene biotechnology*, 3rd ed., Himalaya Publishing House, New Delhi, pp. 171–186.

John, R. P., Gangadharan, D., Nampoothiri, K. M. (2008). Genome shuffling of *Lactobacillus delbueckii* mutant and *Baccillus amyloliquefaciens* through protoplasmic fusion for L-lactic production from starchy wastes. *Biores. Technol.* 99, 8008–8015.

Kapoor, M. (1995). Gene transfer by electroporation of filamentous fungi. *In Electroporation protocols for microorganisms*, ed. by Nickoloff, J. K., Springer, New York, pp. 279–289.

Klein, T. M., Wolf, E. D., Wu, R., Sanford, J. C. (1987). High velocity microprojectiles for delivering nucleic acids into living cells. *Nature* 327, 70–73.

Kück, U., Hoff, B. (2010). New tools for the genetic manipulation of filamentous fungi. *Appl Microbiol Biotechnol.* 86(1), 51–62.

Kumar, M. (2007). Improving polycyclic aromatic hydrocarbons degradation by genome shuffling. *Asian J. Microbiol. Biotechnol. Environ. Sci.* 9, 145–149.

Kwon-Chung, K. J., Goldman, W. F., Klein, B., Szaniszlo, P. J. (1998). Fate of transforming DNA in pathogenic fungi. *Medical Mycology* 36, 38–44.

Legmann, R., Margalith, P. (1983). Interspecific protoplast fusion of *Saccharomyces cerevisiae* and *Saccharomyces mellis*. *Eur J Appl Microbiol Biotechnol.* 18, 320–322.

Lein, J. (1986). The Panlabs penicillin strain improvement program. *In Over production of Microbial Metabolites; Strain Improvement and Process Control Strategies*, ed. by Vanek, Z., Hostalek, Z., Butterworth Publishers, Boston, MA, pp. 105–139.

Leja, K., Myszka, K., Czaczyk, K. (2011). Genome shuffling: a method to improve biotechnological processes. *BioTechnologia* 92(4), 345–351.

Liang, H. Y., Guo, Y. (2007). Whole genome shuffling to enhance activity of fibrinolytic enzyme producing strains. *China Biotechnol.* 27, 39–43.

Lin, J., Shi, B. H., Shi, Q. Q., He, Y. X., Wang, M. Z. (2007). Rapid improvement in lipase production of *Penicillium expansum* by genome shuffling. *Chin. J. Biotechnol.* 23(4), 672–676.

Lorito, M., Hayes, C. K., Pietro, A. D., Harman, G. E. (1993). Biolistic transformation of *Trichoderma harzianum* and *Gliocladium virens* using plasmid and genomic DNA. *Curr. Genet.* 24, 349–356.

Mach, R. L., Zeilinger, S. (2003). Regulation of gene expression in industrial fungi: *Trichoderma*. *Appl Microbiol Biotechnol.* 60, 515–522.

MacKenzie, D. A., Jeenes, D. J., Archer, D. B. (2004). Filamentous fungi as expression systems for heterologous proteins. In *The Mycota II. Genetics and Biotechnology*, ed. by Kuck, U., Springer-Verlag, New York, pp. 289–315.

Maras, M., van Die, I., Contreras, R., van den Hondel, C. A. M. J. J. (1999). Filamentous fungi as production organisms for glycoproteins of bio-medical interest. *Glycoconj J.* 16, 99–107.

Meyer, H. P., Biass, J., Jungo, C., Klein, J., Wenger, J., Mommers, R. (2008). An emerging star for therapeutic and catalytic protein production. *BioProc Internat.* 6, 10–21.

Meyer, V., Mueller, D., Strowig, T., Stahl, U. (2003). Comparison of different transformation methods for *Aspergillus giganteus*. *Curr Genet.* 43, 371–377.

Michielse, C. B., Hooykaas, P. J. J., C. A. M. J. J. van den Hondel, Ram, A. F. J. (2005). *Agrobacterium*-mediated transformation as a tool for functional genomics in fungi. *Curr. Genet.* 48, 1–17.

Murlidhar, R. V., Panda, T. (2000). Fungal protoplast fusion: a revisit. *Bioprocess Biosyst Engg.* 22, 429–431.

Narayanswamy, S. (1994). *Plant cells and tissue cultures, plant protoplast: isolation, culture and fusion*, TATA-McGraw Hill Publishing Company, New Delhi, pp. 391–469.

Nazari, R., Akbarzadeh, A., Norouzian, D., Farahmand, B., Vaez, J., Sadegi, A., Hormozi, F., Rad, K. M., Zarbaksh, B. (2005). Applying intraspecific protoplast fusion in *Streptomyces griesoflavus* to increase the production of Desferrioxamines B. *Curr Sci.* 88(11), 1815–1820.

Okanishi, M., Suzuki, N., Furuta, T. (1996). Variety of hybrid characters among recombinants obtained by interspecific protoplast fusion in streptomycetes. *Biosci Biotechnol Biochem* 60, 1233–1238.

Olempska-Beer, Z. S., Merker, R. I., Ditto, M. D., Di Novi, M. J. (2006). Food-processing enzymes from recombinant microorganisms – a review. *Regul Toxicol Pharmacol.* 45, 144–158.

Otte, B., Grunwaldt, E., Mahmoud, O., Jennewein, S. (2009). Genome shuffling in *Clostridium diolis* DSM 15410 for improved 1,3-propanediol production. *Appl. Environ. Microbiol.* 75, 7610–7616.

Ozeki, K., Kyoya, F., Hizume, K., Kanda, A., Hamachi, M., Nunokawa, Y. (1994). Transformation of intact *Aspergillus niger* by electroporation. *Biosci. Biotechnol. Biochem.* 58, 2224–2227.

Panchal, C. J., Harbison, A., Russell, I., Stewart, G. G. (1982). Ethanol production of genetically modified strains of *Saccharomyces*. *Biotechnol Lett* 4, 33–38.

Park, S. M., Kim, D. K. (2004). Transformation of a filamentous fungus *Cryphonectria parasitica* using *Agrobacterium tumefaciens*. *Biotechnol Bioprocess Eng.* 9, 217–222.

Patnaik, R., Louie, S., Gavrilovic, V., Perry, K., Stemmer, W. P., Ryan, C. M., del Cardayré, S. (2002). Genome shuffling of *Lactobacillus* for improved acid tolerance. *Nat. Biotechnol.* 20, 707–712.

Petri, R., Schmidt-Dannert, C. (2004). Dealing with complexity: evolutionary engineering and genome shuffling. *Curr. Opin. Biotechnol.* 15, 298–304.

Piers, K. L., Heath, J. D., Liang, X., Stephens, K. M., Nester, E. W. (1996). *Agrobacterium tumefaciens*-mediated transformation of yeast. *Proc Natl Acad Sci USA.* 93, 1613–1618.

Polizeli, M. L., Rizzatti, A. C., Monti, R., Terenzi, H. F., Jorge, J. A., Amorim, D. S. (2005). Xylanases from fungi: properties and industrial applications. *Appl Microbiol Biotechnol.* 67, 577–591.

Powell, K. A., Ramer, S. W., Cardayre, S. B. D., Stemmer, W. P., Tobin, M. B., Longchamp, P. F., Huisman, G. W. (2001). Directed evolution and biocatalysis. *Angew. Chem. Int. Ed. Engl.* 40(21), 3948–3959.

Prabavathy, V. R., Mathivanan, N., Sagadevan, E., Murugesan, K., Lalithakumari, D. (2006). Intra-strain protoplast fusion enhances carboxymethyl cellulase activity in *Trichoderma reesei*. *Enzyme Microb. Technol.* 3, 719–723.

Prabha, V. L., Punekar, N. S. (2004). Genetic transformation in *Aspergilli*: tools of the trade. *Indian Journal of Biochemistry and Biophysics* 42, 205–215.

Primrose, S. B., Twyman, R. M. (2006). *Principles of gene manipulation and genomics*, Blackwell Publishing, Oxford, UK.

Punt, P. J., van Biezen, N., Conesa, A., Albers, A., Mangnus, J., van den Hondel, C. A. (2002). Filamentous fungi as cell factories for heterologous protein production. *Trends Biotechnol.* 20, 200–206.

Riach, M. B. R., Kinghorn, J. R. (1996). Genetic transformation and vector developments in filamentous fungi. *In Fungal Genetics: Principles and Practice*, ed. by Bos, C. J., Marcel Dekker Inc., New York, pp. 209–233.

Richey, M. G., Marek, E. T., Schard, C. L., Smith, D. A. (1989). Transformation of filamentous fungi with plasmid DNA by electroporation. *Phytopathology* 79(8), 844–847.

Ruiz-Diez, B. (2002). Strategies for the transformation of filamentous fungi. *Journal of Applied Microbiology* 92, 189–195.

Ryu, D. D. Y., Kim, K. S., Cho, N. Y., Pai, H. S. (1983). Genetic recombination in *Micromonospora rosaria* by protoplast fusion. *Appl Environ Microbiol.* 45, 1854–1858.

Sánchez, O., Aguirre, J. (1996). Efficient transformation of *Aspergillus nidulans* by electroporation of germinated conidia. *Fungal Genetics Newsletter* 43, 48–51.

Schwab, H. (1988). Strain improvement in industrial microorganisms by recombinant DNA techniques. *Advances in Biochemical Engineering/Biotechnology* 37, 129–168.

Shi, D. J., Wang, C. L., Wang, K. M. (2009). Genome shuffling to improve thermotolerance, ethanol tolerance and ethanol productivity of *Saccharomyces cerevisiae*. *J. Ind. Microbiol. Biotechnol.* 36, 139–147.

Shi, J., Zhang, M., Zhang, L., Wang, P., Jiang, L., Deng, H. (2014). Xylose-fermenting *Pichia stipitis* by genome shuffling for improved ethanol production. *Microbial Biotechnology* 7(2), 90–99.

Skatrud, P. L., Tietz, A. J., Ingolia, T. D., Cantwell, C. A., Fisher, D. L., Chapman, J. L., Queener, S. W. (1989). Use of recombinant DNA to improve production of Cephalosporin C by *Cephalosporium acremonium*. *Nature Biotechnology* 7, 477–485.

Srinivas, R., Panda, T. (1997). Localization of carboxymethyl cellulase in the intergeneric fusants of *Trichodermma reesei* QM 9414 and *Saccharomyces cerevisee* NCIM (3288). *Bioprocess Biosystems Engg.* 18, 71–73.

Stemmer, W. P. C. (2002). Molecular breeding of genes, pathway and genomes by DNA shuffling. *J Mol Catal B: Enzym.* 20, 3–12.

Traxler, P., Schupp, T., Wehrli, W. (1982). 16,17-dihydrorifamycin S and 16,17- dihydro-17-hydroxyrifamycin S, two novel rifamycins from a recombinant strain C5/42 of *Nocardia mediterranei*. *J Antibiot.* 35, 594–601.

Vijn, I., Govers, F. (2003). *Agrobacterium tumefaciens* mediated transformation of the oomycete plant pathogen *Phytophthora infestans*. *Mol Plant Pathol.* 4, 459–467.

Wang, L., Ridgway, D., Gu, T., Moo-Young, M. (2005). Bioprocessing strategies to improve heterologous protein production in filamentaous fungi. *Biotechnology Advances* 23, 115–129.

Weaver, J. C. (1995). Electroporation theory: concepts and mechanism. In *Electroporation protocols for microorganisms*, ed. by Nickoloff, J. A., Springer, Berlin, Heidelberg, pp. 1–26.

Wesseling, A. C., Lago, B. (1981). Strain improvement by genetic recombination of cephamycin producers, *Nocardia lactamdurans* and *Streptomyces griseus*. *Devel Indust Microbiol.* 22, 641–651.

Wiebe, M. G. (2003). Stable production of recombinant proteins in filamentous fungi – problems and improvements. *Mycologist* 17, 140–144.

Willke, T., Vorlop, K. D. (2001). Biotechnological production of itaconic acid. *Appl Microbiol Biotechnol.* 56, 289–295.

Xu, B., Jin, Z. H., Wang, H. Z., Jin, Q. C., Jin, X., Cen, P. L. (2008). Evolution of *Streptomyces pristinaespiralis* for resistance and production of pristinamycin by genome shuffling. *Appl. Microbiol. Biotechnol.* 80, 261–267.

Zhang, Y., Liu, J. Z., Huang, J. S., Mao, Z. M. (2010). Genome shuffling of *Propionibacterium shermanii* for improving vitamin B12 production and comparative proteome analysis. *J. Biotechnol.* 148, 139–143.

Zhang, Z. B., Burgos, N. R., Zhang, J. P., Yu, L. O. (2007). Biological control agent for rice weeds from protoplast fusion between *Curvularia lunata* and *Helminthosporium gramineum. Weed Sci.* 55, 599–605.

Zhao, K., Ping, W. X., Zhang, L., Liu, J., Lin, Y., Jin, T., Zhou, D. (2008). Screening and breeding of high taxol producing fungi by genome shuffling. *Science in China Series C: Life Sciences* 51(3), 222–231.

CHAPTER 7

BIOINFORMATIC APPROACHES FOR IDENTIFICATION OF FUNGAL SPECIES AND DETECTION OF SECONDARY METABOLITES

CONTENTS

7.1   INTRODUCTION

When it comes to molecular identification, fungi are considered as an organism where technology has outpaced taxonomy (Sangeetha and Thangadurai, 2013a). The unexceptional characteristics of fungi coupled with the absence of reliable morphological characters for identification and to differentiate from other species have made it difficult for a comprehensive understanding of fungal kingdom (Sangeetha and Thangadurai, 2013b). DNA sequence analysis has emerged as a predominant means of taxonomic identification of species, specifically that are minute or lacking distinct, observable morphological features like fungi (Slippers et al., 2013). A number of successful attempts are being made and some are underway to develop standardized protocols and tools for fungal analysis (Gupta et al., 2013). Initially most researchers either developed their own set of tools or failed to distinguish due to lack of tools (Sangeetha and Thangadurai, 2013b). Recent advancements in bioinformatics have made the fungal isolation, identification, classification, and phylogenetic analysis an uncomplicated task and are increasing day by day (Rastogi and Sani, 2011). The current dramatic increase in the number and scale of genome sequence and functional genomic data in a similar way to proteomic, microarray and RNA sequence has made it more and more demanding for scientists to navigate through the substantial amount of data (Matsuzaki et al., 2004; Galagan et al., 2005; Derelle et al., 2006; Gregory et al., 2006; Tuskan et al., 2006). Despite the hard efforts from bioinformatics society, large quantity of genomes continues to remain unannotated. The present situation demands the need of a simple, user-friendly software tool. Therefore, it is important to utilize bioinformatics databases and tools, which are organized and analyzed these vast amounts of information. There are good number of fungal databases that were developed as a resource for structural genomic and functional genomic data across the fungal kingdom.

7.2   CANDIDA GENOME DATABASE

The Candida Genome Database (CGD) is a public resource containing genomic information for those who are interested in the molecular biology studies of fungal pathogen, *Candida albicans*. Researchers in CGD are collecting and combining previous research work to collect *Candida albicans* gene name and aliases to assign gene ontology terms, which provide the information regarding the molecular function, biological process and subcellular localization of each gene product. In addition, CGD is used to annotate mutant phenotypes and to summarize the function and biological context of each gene product in free-text description lines. The CGD search tools, such as Quick Search, Text Search, Gene/Sequence Resources, Ortholog Search and Pattern Match are designed according to search multiple species. Search results displayed on the top for all species, with sections for species-specific search results displayed below. CGD tools help to perform species- or sequence-specific searches (e.g., Gene/Sequence Resources, Pattern Match, Advanced Search, Batch Download, Restriction Mapper, GO Term Finder, GO Slim Mapper). The CGD Locus Summary Page (LSP) provides information about the identity of orthologous genes and orthology-based functional predictions and gene descriptions in *Candida glabrata*. Both manual and computational gene, protein and sequence information of *Candida albicans* and the recently added species, *Candida glabrata* are displayed under CGD (Inglis et al., 2012). BLAST searches at CGD provide complete sequence sets for the combination of several *Candida* species, such as *Candida albicans*, *Candida glabrata*, *Candida dubliniensis*, *Candida guilliermondii*, *Candida lusitaniae*, *Candida parapsilosis*, *Candida tropicalis*, and *Debaryomyces hansenii* and *Lodderomyces elongisporus* (Altschul et al., 1990; Jones et al., 2004; Dujon et al., 2004; Butler et al., 2009). LSP is the central organizing unit of the CGD website and it represents each gene in CGD. The LSP provides access to tools for retrieval, analysis and visualization of gene data. It also acts as a platform to access information about the orthologs in *Saccharomyces cerevisiae* (Inglis et al., 2012). InParanoid algorithm is used to define orthology relationships, which identify reciprocal best BLAST hits between species (Remm et al., 2001). Protein tab on LSP provides similarity based information of each protein-coding gene through descriptions and a graphical display of conserved protein domains and motifs identified using InterProScan software (Zdobnov and Apweiler, 2001; Hunter et al., 2009). It displays the most similar protein in the Protein Data Bank and provides information on protein length, molecular weight, sequence and a table of calculated physicochemical properties (Rose et al., 2010). This is publicly funded and is freely available at http://www.candidagenome.org (Arnaud et al., 2005).

7.3 COMPREHENSIVE YEAST GENOME DATABASE

The Comprehensive Yeast Genome Database (CYGD) provides information on the molecular structure, metabolic regulatory pathways, signal transduction, transport process of co-regulated gene clusters and functional network of the entirely sequenced, well-studied yeast *Saccharomyces cerevisiae*. In addition, for comparative analysis the data of various projects on related yeasts are used. More yeast genomes annotation greatly facilitated using *Saccharomyces cerevisiae* as a reference. Population of CYGD catalogs are FunCat, EC, protein classes, protein complexes, localization, phenotypes and transporter. CYGD, with the aid of *Saccharomyces cerevisiae* genome as a backbone and SIMAP (SImilarity MAtrix of Protein Sequence), provides a way for exploring related genomes (Ruepp et al., 2004; Güldener et al., 2005). The SIMAP sequences provides pre-calculated comparison of the protein sets of all genomes analyzed by PEDANT, as well as from other sources like Swiss-Prot (Frishman et al., 2000; Frishman et al., 2003) using FASTA package (Pearson, 2000). The entire structure of the databases allows the annotation of complex relationships such as protein-protein interactions using GenRE. The comprehensive resource is available at http://mips.gsf.de/genre/proj/yeast/ (Guldener et al., 2005).

7.4 FUNGAL CYTOCHROME P450 DATABASE

Collection of vast genome will always interrupt the data analysis and therefore, fungal cytochrome P450 database (FCPD) concentrates on particular enzyme coding for cytochrome P450 (Park et al., 2008a; Moktali et al., 2012). It is an essential enzyme in the degradation of organic substances, which belongs to heme protein. From 66 fungal and 4 oomycete genomes, 4538 genes have been maintained in FCPD that codes for cytochrome P450 enzyme. This database also provides information about how to analyze the sequences related to P450 enzyme, evolutionary relationship between the species and gives detailed information about chromosomal distribution. Information collected on particular enzyme is further classified into 16 classes, 141 clustered groups and domain information using InterPro database by applying tribe-MCL tool (Enright et al., 2002). Minimum length of 25 amino acid sequences was used to filter very short domain and these were labeled as questionable P450 instead of discarding. This data reveals that distribution of P450s in species of each depends on taxon character. To avoid the confusion and to speed up the data, authors have maintained cache tables, which are obtained from the result of several sequence analysis. This database also maintains BLAST datasets to search P450s using BLAST and also to analyze the cluster using tribe-MCL (Doğan and Karaçalı 2013). Phylogenetic trees are constructed for specific class and cluster of P450 using Phyloviewer on the web (http://www.phyloviewer.org/). Data obtained was compared with publicly available databases and stored in FCPD. The FCPD can easily be accessed at http://p450.riceblast.snu.ac.kr/ (Park et al., 2008b; Moktali et al., 2012).

7.5   FUNGAL SECRETOME KNOWLEDGEBASE

Fungal Secretome KnowledgeBase (FunSecKB) is the database useful in collection of secreted fungal proteins called secretomes (Choi et al., 2010; Lum and Min, 2011). All the related protein-coding genes are identified from a NCBI RefSeq database (Pruitt and Maglott, 2001; Robbertse and Tatusova, 2011; Pruitt et al., 2012; Tatusova et al., 2014). SignalP, WolfPsort and Phobius are the computational protocol for signal peptide and subcellular location prediction. Tools like TMHMM is used for identifying membrane proteins, if transmembrane domain is located within the N-terminus 70 amino acids, and a signal peptide is predicted by SignalP 4.0, which is not counted as a membrane protein (Otávio et al., 2009; Min, 2010; Meinken and Min, 2012). PS-Scan is used for identifying endoplasmic reticulum target proteins (Castro et al., 2006). TargetP is classified as a chloroplast protein and mitochondrial membrane protein (Emanuelsson et al., 2000; Emanuelsson et al., 2007). GPI-anchored proteins are the signal peptide containing proteins that were predicted by FragAnchor and these proteins are involved in signaling, adhesion, stress response, cell wall remodeling and play different roles in cell growth and development (Poisson et al., 2007). For the fungal protein sequences, which are downloaded from the NCBI RefSeq database, the default parameters for fungi were used for each of these programs. From UniProt and Swiss-Prot data set, 241 secreted proteins and 5992 non-secreted proteins of fungi are experimentally identified using accuracy evaluation of the computational methods (Wu et al., 2006; Bairoch et al., 2008). The highest prediction accuracy was achieved by combining SignalP, WolfPsort and Phobius for signal peptide prediction. For SignalP, SignalP-NN algorithm is used and for signal peptide SignalP-HMM algorithm is used as available at http://proteomics.ysu.edu/secretomes/fungi.ph (Lum and Min, 2011).

7.6   FUNGAL TRANSCRIPTION FACTOR DATABASE

Fungal Transcription Factor Database (FTFD) contains 31,832 putative transcription factors (TFs) from 62 fungal and 3 Oomycete species. They were phylogenetically classified and analyzed into 61 families. Within and across species, the database serves as a community resource for the comparative analysis of distribution and domain structure of TFs using Transcription Factor Matrices (TFM), a software suite for identifying and analyzing transcription factor binding sites modeled by position weight matrices and position specific scoring matrices. A program termed TFM-Explorer (Transcription Factor Matrix Explorer) was developed for analyzing regulatory regions in eukaryotic genomes, as there is a growing interest in deciphering regulatory regions in DNA sequences to understand transcriptional regulation. The program considers a set of coregulated gene sequences to search for locally overrepresented transcription factor binding sites in two steps and further scan sequences for potential transcription factor binding sites using weight matrices by calculating a score function. TFM-Explorer allows to visualize result clusters and to search for cis-regulatory modules. TFM-Scan is an another program dedicated to locate large sets of putative transcription factor binding sites on a DNA sequence using position weight matrices such as those in

Transfac or Jaspar databases. TFM-Scan is also capable to cluster similar matrices and occurrences. The algorithm of FTFD database available at http://ftfd.snu.ac.kr (Park et al., 2008a, b) is very fast and hence allows for large-scale analysis.

7.7 FUNGIDB

FungiDB is an integrated functional genomic database that belongs to the EuPathDB family for the kingdom Fungi. The database contains genomes of 46 fungi and 6 oomycetes including *Aspergillus clavatus*, *Aspergillus flavus*, *Aspergillus fumigatus*, *Aspergillus niger*, *Aspergillus terreus*, *Candida albicans*, *Coccidoidies immitis* strain H538.4, *Coccidoidies immitis* strain RS, *Cryptococcus neoformans* var. *grubii*, *Emericella nidulans*, *Fusarium graminearum*, *Fusarium oxysporum* f. sp. *lycopersici*, *Fusarium verticillioides*, *Magnaporthe grisea*, *Neurospora crassa* strain OR74A, *Puccinia graminis*, *Rhizopus oryzae* and *Saccharomyces cerevisiae*. The specific details about these genomes have been incorporated in this database through a web interface to provide convenient and straightforward access to the available data. It integrates whole genome sequence analysis, gene expression studies, comparative genomics, annotation, data-mining and other supplemental bioinformatics through web interface. Interestingly, the database contains searches and tools that return sets of genes, expressed sequence tags (ESTs), open reading frames (ORFs), genomic segments (including DNA motifs), genomic sequences, GMOD genome browser, BLAST, sequence retrieval and fungal-related literature purely based on PubMed searches. FungiDB is freely available for registered users at http://fungidb.org/fungidb/ (Stajich et al., 2012).

7.8 FUNYBASE

FUNYBASE (FUNgal phYlogenomic dataBASE) is one of the recent databases which helps in maintaining the information of homologous sequence of common ancestor of fungal genome to avoid confusion of heterogeneity of data and this database provide comparative and phylogenetic analyzes that is constructed by using amino acid sequences in fungi. FUNYBASE serve as foremost resource for study of fungal comparative genomics, as it concedes the retrieval of ortholog clusters, thereby representing the major fungal taxonomic groups across a large phylogenetic scale. It is implemented on the relational database system PostgreSQL (version 8.2.4) and this data can aid in multiple purposes including gene comparison, gene searching and tree comparison. Currently, FUNYBASE has two types of protein data. Firstly, clusters of orthologs are created and classified by using publicly available information of more than 30 fungal genome protein sequences by automated procedures. Secondly, subsets of 246 ortholog clusters as single copy genes have been recovered from 21 fungal genomes (Aguileta et al., 2008). Based on the aminoacid sequences phylogenetic tree is reconstructed for each of these ortholog clusters and the classification of these ortholog clusters depends on the topological score, which make use of WAG evolutionary model. By using WAG evolutionary model, species tree are constructed by using a series of approximately half of the 246 sequences. To assess the informative value of every ortholog cluster, the same is compared to a reference species tree. Suppose if it measures the same topology score then it is referred as same species tree. These results are available on-line which enables search for species name, the ortholog cluster and various keywords. This is achieved by making use of the BLAST algorithm. A basic local alignment search tool for proteins (BLASTP) search of each predicted protein sequences against the entire assembled protein sequences database was performed using the NCBI BLAST2 software (Abascal et al., 2005). Currently, the database is available at http://genome.jouy.inra.fr/funybase/ (Marthey et al., 2008).

7.9   FUSARIUM GRAMINEARUM GENOME DATABASE

Fusarium graminearum, a plant pathogen which mainly causes head blight disease to wheat, barley and some other crop species, produces mycotoxins that will affect the health of animals and human beings (Trail, 2009). Identification of this pathogenic fungus is very important and *Fusarium graminearum* genome database (FGDB) involves in the identification of toxic gene by using bioinformatics tools based on molecular interaction network and gene expression data (Liu et al., 2010; Becher et al., 2011; Wong et al., 2011). By using subnetwork which contained seed genes of known pathogenic fungi was made to anneal with *Fusarium graminearum*. The predicted result showed that most of the pathogenic genes of *Fusarium graminearum* are enriched in two important signal transduction pathways, including G protein coupled receptor pathway and MAPK signaling pathway (Liu et al., 2010). A set of transcription-associated proteins (TAPs) was taken from the UniProt database (Bairoch et al., 2008). Using TRIBE-MCL, *Fusarium graminearum* TAPs were identified by using homologous sequence of *Fusarium* and their matching reference-set (Lawler et al., 2013). Further, TAPs were clustered and the resulted value of 2.0 will ensure maximum coverage of the corresponding protein families. Once detected these protein families were placed into one of the five TAP categories. DNA binding domains present in *Fusarium graminearum* genome can be identified by using Hidden-Markov models (HMMs). This database provides search options on the sidebar for gene codes, gene symbols, gene description and the annotation catalogs (Srivastava et al., 2014). FunCat, Enzyme Class, InterPro and Protein Class are browsable and the details of data, such as probe, probe set names, their location on tRNAs and a customizable on protein molecular weights with isoelectric points, have been made available in the database. The ORF or contig DNA and protein sequences are searchable by Blast. SIMAP based protein homology data can be retrieved using links grouped by NCBI-based taxonomic categories and the most recent expression data is provided by a link to the 'PLEXdb GeneOscilloScope' at http://mips.gsf.de/genre/proj/fusarium/ (Bairoch et al., 2008).

7.10   INTERNATIONAL NUCLEOTIDE SEQUENCE DATABASE

International Nucleotide Sequence Database (INSD) is the database formed by the collaboration of three organizations consisting of DNA Data Bank of Japan (DDBJ) located at National Institute of Genetics (NIG), European Molecular Biology Laboratory (EMBL) at European Bioinformatics Institute and GenBank at the National Center for Biotechnology Information. Biological database that collect DNA sequences remains the same at each organization as information is shared on daily basis. INSD deposited data helps to identify mycorrhizal fungi using ITS sequence analysis. These sequences are downloaded from INSD through BLAST, sequence are aligned and examined by phylogenetic tree. INSD contains ITS sequence of 83,208 related to roots of plants and orchids including 28,791 and 3,176 sequences that are related to EcM fungi and AM fungi respectively. Comparison study of ITS1 and ITS2 for DNA metabarcodes in fungi was made by collecting the information about ITS obtained from INSD. The database is available at http://www.insdc.org/ (Brunak et al., 2002; Barrett et al., 2011; Cochrane et al., 2011; Parkinson et al., 2011; Barrett et al., 2012; Karsch-Mizrachi et al., 2012; Kodama et al., 2012).

7.11   Q-BANK FUNGI DATABASE

A group of mycologists from Netherlands have contributed for the construction of Q-bank database, in which bulk of DNA sequence (barcode) data of fungi is updated regularly. Q-bank fungi database contains data regarding more than 725 species of fungi that are related to phytopathology, saprophytic organisms and information about morphological, phenotypical and ecological data. On contemporary the database focuses on members, especially those of quarantine importance to Europe and their closest relatives, of the fungal genera *Phoma* and associated genera such as *Colletotrichum, Ceratocystis, Monilinia, Mycosphaerella* and its anamorphs, *Puccinia, Stenocarpella, Thecaphora, Verticillium,* and the Oomycete genus *Phytophthora.* This database helps in correct identification of the causal agent of diseases by comparing between different organisms through sequence information in public online databases. Q-bank fungi contain both species and strains information followed by the link between the species and strains. The available fungal strain in Q-bank database are ordered by strain specific external links, such as culture collections or contact information of the strain. Unknown identification of a fungal species starts with a "Blast against all Q-bank (fungal) sequences" from fungal Q-bank page. Any non-standard symbols or alignment gap should be present in the sequence. By clicking on the prime hyperlink in the hits retrieved from the search in opening the page that has the comprehensive data of the entry. The identification of species of interest to which fungal organism groups it belongs can be found out by using the item "organism group," that is present at the top of web page. With this information a focused multilocus search can be executed for specific fungal group under the heading of "ID," and the identification of multilocus "organism group" can be performed under the menu item "multilocus sequence ID per group." Within this there are two tabs, one is "description" wherein, the title or description is entered in 'item name' about the analysis to be performed and secondly, "sequences" in which the sequences are pasted into the available corresponding locus blocks; then the search can be conducted by clicking on the "start identification" object that is present at the bottom of page. After the search is completed the results are obtained as percentage similarity or viewed as plylogenetic tree by using "draw tree" item. The Q-bank fungi database can be accessed through http://www.q-bank.eu/ (Quaedvlieg et al., 2012; Tanabe and Toju, 2013; van de Vossenberg et al., 2013).

7.12 UNITE

UNITE is an rDNA sequence database which holds only sequences from the ITS region of ectomycorrhizal fungi (Bruns et al., 2008). ITS sequences are widely used for the identification fungi as it helps in discriminating between closely related fungal species. The generated sequences were from fruit bodies that are available in public herbaria and type specimens (Kõljalg et al., 2005). UNITE holds more vital information on herbarium, geographical location, morphological description of mycelia, fruiting-body, illustrations, ecology, taxonomy and nomenclature, that are essential tools for interpretation. Full descriptions and illustrations for selected species were also linked to the sequences. Until now, UNITE has 758 ITS sequences from 455 species and 67 genera of basidiomycetes and ascomycetes (Abarenkov et al., 2010). Several tools were incorporated into the UNITE database, and together, they aid in the identification of unknown sequences beyond simple similarity searches. For example, two galaxie tools namely GalaxieBLAST and GalaxieHMM have been implemented for web-based basic phylogenetic analyzes (Nilsson et al., 2004).

These galaxie tools are most appropriate for identification of unknown ITS sequences which provide maximum parsimony heuristic and neighbor-joining analyzes under different evolutionary models (Thompson et al., 1994). The most recent version, a stable and reliable platform of UNITE 5.0 is available at http://unite.ut.ee/ (Pukkila and Skrzynia, 1993; Sonnenberg et al., 1996; Altschul et al., 1997; Eddy, 1998; Eberhardt et al., 1999; Lilleskov et al., 2002; Moncalvo et al., 2002; Vrålstad et al., 2002; Kjøller and Bruns, 2003; Rosling et al., 2003; Tautz et al., 2003; Tedersoo et al., 2003; Larsson et al., 2004; Parmasto et al., 2004; Will and Rubinoff, 2004; Kõljalg et al., 2013).

KEYWORDS

- Annotation
- *Aspergillus clavatus*
- *Aspergillus flavus*
- *Aspergillus fumigatus*
- *Aspergillus niger*
- *Aspergillus terreus*
- Bioinformatics databases
- *Candida albicans*
- Candida Genome Database
- CGD Locus Summary Page
- CGD search tools
- Chloroplast protein
- *Cis*-regulatory modules
- *Coccidoidies immitis*
- Comparative genomics
- Comprehensive Yeast Genome Database
- Cytochrome P450
- Data-mining
- DNA binding domains

- **DNA Data Bank of Japan**
- **DNA metabarcodes**
- **DNA motifs**
- **Ectomycorrhizal fungi**
- **Endoplasmic reticulum target proteins**
- **EuPathDB**
- **Expressed sequence tags**
- **FASTA package**
- **FragAnchor**
- **FunCat**
- **Fungal cytochrome P450 database**
- **Fungal proteins**
- **Fungal Secretome KnowledgeBase**
- **Fungal Transcription Factor Database**
- **FungiDB**
- **FUNYBASE**
- ***Fusarium graminearum***
- ***Fusarium graminearum* genome database**
- **GalaxieBLAST**
- **GalaxieHMM**
- **Gene expression**
- **Genome sequence**
- **Genomic segments**
- **GenRE**
- **GMOD genome browser**
- **GPI-anchored proteins**
- **Head blight disease**
- **Heme protein**
- **Hidden-Markov models**
- **Homologous sequence**
- **International Nucleotide Sequence Database**

- **InterPro database**
- **InterProScan software**
- **Jaspar databases**
- **MAPK signaling pathway**
- **Maximum parsimony**
- **Membrane proteins**
- **Metabolic regulatory pathways**
- **Microarray**
- **Mitochondrial membrane protein**
- **National Center for Biotechnology Information**
- **NCBI RefSeq database**
- **Neighbour-joining**
- **Open reading frames**
- **Ortholog clusters**
- **Paranoid algorithm**
- **Phobius**
- **Phylogenetic analysis**
- **Phylogenetic trees**
- **Phyloviewer**
- **PLEXdb GeneOscilloScope**
- **Position specific scoring matrices**
- **Position weight matrices**
- **PostgreSQL**
- **Protein coding genes**
- **Protein Data Bank**
- **Protein-protein interactions**
- **PS-Scan**
- **Q-bank Fungi Database**
- **rDNA sequence database**
- **RefSeq database**
- ***Saccharomyces cerevisiae***
- **Secretomes**

- **Signal peptide**
- **Signal transduction**
- **SignalP-HMM algorithm**
- **SignalP-NN algorithm**
- **Subcellular location prediction**
- **Swiss-Prot data set**
- **TargetP**
- **TFM-Explorer**
- **TFM-Scan**
- **Transcription factors**
- **Transcription-associated proteins**
- **Transfac**
- **Tribe-MCL tool**
- **UniProt database**
- **UNITE database**
- **WAG evolutionary model**
- **Whole genome sequence analysis**
- **WolfPsort**

REFERENCES

Abarenkov, K., Nilsson, R. H., Larsson, H. K., Alexander, I. J., Eberhardt, U., Erland, S., Høiland, K., Kjøller, R., Larsson, E., Pennanen, T., Sen, R., Taylor, A. F. S., Tedersoo, L., Ursing, B. M., Vrålstad, T., Liimatainen, K., Peintner, U., Kõljalg, U. (2010). The UNITE database for molecular identification of fungi – recent updates and future perspectives. *New Phytologist* 186(2), 281–285.

Abascal, F., Zardoya, R., Posada, D. (2005). ProtTest: selection of best-fit models of protein evolution. *Bioinformatics* 21, 2104–2105.

Aguileta, G., Marthey, S., Chiapello, H., M.-Lebrun, H., Rodolphe, F., Fournier, E., Gendrault-Jacquemard, A., Giraud, T. (2008). Assessing the performance of single-copy genes for recovering robust phylogenies. *Syst Biol.* 57, 613–627.

Altschul, S. F., Madden, T. L., Schaffer, A. A., Zhang, J., Zhang, Z., Miller, W., Lipman, D. J. (1997). Gapped blast and PSI-blast: a new generation of protein database search programs. *Nucl. Acids Res.* 25, 3389–3402.

Arnaud, M. B., Costanzo, M. C., Skrzypek, M. S., Binkley, G., Lane, C., Miyasato, S. R., Sherlock, G. (2005). The Candida Genome Database (CGD), a community resource for *Candida albicans* gene and protein information. *Nucl. Acids Res*. 33, D358–D363.

Bairoch, A., Bougueleret, L., Altairac, S., Amendolia, V., Auchincloss, A., Puy, G. A., Axelsen, K., Baratin, D., Blatter, M. C., Boeckmann, B., Bollondi, L., Boutet, E., Quintaje, S. B., Breuza, L., Bridge, A., Bulliard-Le Saux, V., de Castro, E., Ciampina, L., Coral, D., Coudert, E., Cusin, I., David, F., Delbard, G., Dornevil, D., Duek-Roggli, P., Duvaud, S., Estreicher, A., Famiglietti, L., Farriol-Mathis, N., Ferro, S., Feuermann, M., Gasteiger, E., Gateau, A., Gehant, S., Gerritsen, V., Gos, A., Gruaz-Gumowski, N., Hinz, U., Hulo, C., Hulo, N., Innocenti, A., James, J., Jain, E., Jimenez, S., Jungo, F., Junker, V., Keller, G., Lachaize, C., Lane-Guermonprez, L., Langendijk-Genevaux, P., Lara, V., Le Mercier, P., Lieberherr, D., Lima, T. O., Mangold, V., Martin, X., Michoud, K., Moinat, M., Morgat, A., Nicolas, M., Paesano, S., Pedruzzi, I., Perret, D., Phan, I., Pilbout, S., Pillet, V., Poux, S., Pozzato, M., Redaschi, N., Reynaud, S., Rivoire, C., Roechert, B., Sapsezian, C., Schneider, M., Sigrist, C., Sonesson, K., Staehli, S., Stutz, A., Sundaram, S., Tognolli, M., Verbregue, L., Veuthey, A. L., Vitorello, C., Yip, L., Zuletta, L. F., Apweiler, R., Alam-Faruque, Y., Barrell, D., Bower, L., Browne, P., Chan, W. M., Daugherty, L., Donate, E. S., Eberhardt, R., Fedotov, A., Foulger, R., Frigerio, G., Garavelli, J., Golin, R., Horne, A., Jacobsen, J., Kleen, M., Kersey, P., Laiho, K., Legge, D., Magrane, M., Martin, M. J., Monteiro, P., O'Donovan, C., Orchard, S., O'Rourke, J., Patient, S., Pruess, M., Sitnov, A., Whitfield, E., Wieser, D., Lin, Q., Rynbeek, M., di Martino, G., Donnelly, M., van Rensburg, P., Wu, C., Arighi, C., Arminski, L., Barker, W., Chen, Y., Crooks, D., Hu, Z. Z., Hua, H. K., Huang, H., Kahsay, R., Mazumder, R., McGarvey, P., Natale, D., Nikolskaya, A. N., Petrova, N., Suzek, B., Vasudevan, S., Vinayaka, C. R., Yeh, L. S., Zhang, J. (2008). The universal protein resource (UniProt). *Nucl. Acids Res*. 36, D190–D195.

Barrett, T., Troup, D. B., Wilhite, S. E., Ledoux, P., Evangelista, C., Kim, I. F., Tomashevsky, M., Marshall, K. A., Phillippy, K. H., Sherman, P. M., Muertter, R. N., Holko, M., Ayanbule, O., Yefanov, A., Soboleva, A. (2011). NCBI GEO: archive for functional genomics data sets – 10 years on. *Nucl. Acids Res*. 39, D1005–D1010.

Barrett, T., Clark, K., Gevorgyan, R., Gorelenkov, V., Gribov, E., Karsch-Mizrachi, I., Kimelman, M., Pruitt, K. D., Resenchuk, S., Tatusova, T., Yaschenko, E., Ostell, J. (2012). BioProject and BioSample databases at NCBI: facilitating capture and organization of metadata. *Nucl. Acids Res*. 40, D57–D63.

Becher, R., Weihmann, F., Deising, H. B., Wirsel, S. G. R. (2011). Development of a novel multiplex DNA microarray for *Fusarium graminearum* and analysis of azole fungicide responses. *BMC Genomics* 12, 52, doi:10.1186/1471-2164-12-52.

Brunak, S., Danchin, A., Hattori, M., Nakamura, H., Shinozaki, K., Matise, T., Preuss, D. (2002). Nucleotide sequence database policies. *Science* 298 (5597), (1333).

Bruns, T. D., Arnold, A. E., Hughes, K. W. (2008). Fungal networks made of humans: UNITE, FESIN, and frontiers in fungal ecology. *New Phytologist* 177, 586–588.

Castro, E., Sigrist, C. J., Gattiker, A., Bulliard, V., Langendijk-Genevaux, P. S., Gasteiger, E., Bairoch, A., Hulo, N. (2006). ScanProsite: detection of PROSITE signature matches and ProRule-associated functional and structural residues in proteins. *Nucl. Acids Res*. 34, W362–365.

Choi, J., Park, J., Kim, D., Jung, K., Kang, S., Lee, Y. H. (2010). Fungal secretome database: integrated platform for annotation of fungal secretomes. *BMC Genomics* 11, 105.

Cochrane, G., Karsch-Mizrachi, I., Nakamura, Y. (2011). The international nucleotide sequence database collaboration. *Nucl. Acids Res.* 39, D15-D18.

Derelle, E., Ferraz, C., Rombauts, S., Rouzé, P., Worden, A. Z., Robbens, S., Partensky, F., Degroeve, S., Echeynié, S., Cooke, R., Saeys, Y., Wuyts, J., Jabbari, K., Bowler, C., Panaud, O., Piégu, B., Ball, S. G., Ral, J. P., Bouget, F. Y., Piganeau, G., de Baets, B., Picard, A., Delseny, M., Demaille, J., Peer, Y., Moreau, H. (2006). Genome analysis of the smallest free-living eukaryote *Ostreococcus tauri* unveils many unique features. *Proc. Natl. Acad. Sci. USA* 103(31), 11647–11652.

Doğan. T., Karaçalı, B. (2013). Automatic identification of highly conserved family regions and relationships in genome wide datasets including remote protein sequences. *PLoS One* 8(9), e75458, doi: 10.1371/journal.pone.0075458.

Eberhardt, U., Walter, L., Kottke, I. (1999). Molecular and morphological discrimination between *Tylospora fibrillosa* and *Tylospora asterophora* mycorrhizae. *Can. J. Bot.* 77, 11–21.

Eddy, S. R. (1998). Profile hidden Markov models. Bioinformatics 14, 755–763.

Emanuelsson, O., Nielsen, H., Brunak, S., Heijne, G. (2000). Predicting subcellular localization of proteins based on their N-terminal amino acid sequence. *J. Mol. Biol.* 300, 1005–1016.

Emanuelsson, O., Brunak, S., Heijne, G., Nielsen, H. (2007). Locating proteins in the cell using TargetP, SignalP and related tools. *Nat. Protoc.* 2, 953–971.

Enright, A. J., van Dongen, S., Ouzounis, C. A. (2002). An efficient algorithm for large-scale detection of protein families. *Nucl. Acids Res.* 30(7), 1575–1584.

Friedman, C. P. (2004). Training the next generation of informaticians: The impact of "BISTI" and bioinformatics – a report from the American College of Medical Informatics. *J. Am. Med. Informatics Assoc.* 11(3), 167–172.

Frishman, D., Albermann, K., Hani, J., Heumann, K., Metanomski, A., Zollner, A., Mewes, H. W. (2000). Functional and structural genomics using PEDANT. *Nucl. Acids Res.* 28(1), 37–40.

Frishman, D., Mokrejs, M., Kosykh, D., Kastenmüller, G., Kolesov, G., Zubrzycki, I., Gruber, C., Geier, B., Kaps, A., Albermann, K., Volz, A., Wagner, C., Fellenberg, M., Heumann, K., Mewes, H. W. (2003). The PEDANT genome database. *Nucl. Acids Res.* 31, 207–211.

Galagan, J. E., Henn, M. R., Ma, L. J., Cuomo, C. A., Birren, B. (2005). Genomics of the fungal kingdom: insights into eukaryotic biology. *Genome Res.* 15, 1620–1631.

Gregory, S. G., Barlow, K. F., McLay, K. E., Kaul, R., Swarbreck, D., Dunham, A., Scott, C. E., Howe, K. L., Woodfine, K., Spencer, C. C., Jones, M. C., Gillson, C., Searle, S., Zhou, Y., Kokocinski, F., McDonald, L., Evans, R., Phillips, K., Atkinson, A., Cooper, R., Jones, C., Hall, R. E., Andrews, T. D., Lloyd, C., Ainscough, R., Almeida, J. P., Ambrose, K. D., Anderson, F., Andrew, R. W., Ashwell, R. I., Aubin, K., Babbage, A. K., Bagguley, C. L., Bailey, J., Beasley, H., Bethel, G., Bird, C. P., Bray-Allen, S., Brown, J. Y., Brown, A. J., Buckley, D., Burton, J., Bye, J., Carder, C., Chapman, J. C., Clark, S. Y., Clarke, G., Clee, C., Cobley, V., Collier, R. E., Corby, N., Coville,

G. J., Davies, J., Deadman, R., Dunn, M., Earthrowl, M., Ellington, A. G., Errington, H., Frankish, A., Frankland, J., French, L., Garner, P., Garnett, J., Gay, L., Ghori, M. R., Gibson, R., Gilby, L. M., Gillett, W., Glithero, R. J., Grafham, D. V., Griffiths, C., Griffiths-Jones, S., Grocock, R., Hammond, S., Harrison, E. S., Hart, E., Haugen, E., Heath, P. D., Holmes, S. S., Holt, K., Howden, P. J., Hunt, A. R., Hunt, S. E., Hunter, G., Isherwood, J., James, R., Johnson, C., Johnson, D., Joy, A., Kay, M., Kershaw, J. K., Kibukawa, M., Kimberley, A. M., King, A., Knights, A. J., Lad, H., Laird, G., Lawlor, S., Leongamornlert, D. A., Lloyd, D. M., Loveland, J., Lovell, J., Lush, M. J., Lyne, R., Martin, S., Mashreghi-Mohammadi, M., Matthews, L., Matthews, N. S., McLaren, S., Milne, S., Mistry, S., Moore, M. J., Nickerson, T., O'Dell, C. N., Oliver, K., Palmeiri, A., Palmer, S. A., Parker, A., Patel, D., Pearce, A. V., Peck, A. I., Pelan, S., Phelps, K., Phillimore, B. J., Plumb, R., Rajan, J., Raymond, C., Rouse, G., Saenphimmachak, C., Sehra, H. K., Sheridan, E., Shownkeen, R., Sims, S., Skuce, C. D., Smith, M., Steward, C., Subramanian, S., Sycamore, N., Tracey, A., Tromans, A., Helmond, Z., Wall, M., Wallis, J. M., White, S., Whitehead, S. L., Wilkinson, J. E., Willey, D. L., Williams, H., Wilming, L., Wray, P. W., Wu, Z., Coulson, A., Vaudin, M., Sulston, J. E., Durbin, R., Hubbard, T., Wooster, R., Dunham, I., Carter, N. P., McVean, G., Ross, M. T., Harrow, J., Olson, M. V., Beck, S., Rogers, J., Bentley, D. R., Banerjee, R., Bryant, S. P., Burford, D. C., Burrill, W. D., Clegg, S. M., Dhami, P., Dovey, O., Faulkner, L. M., Gribble, S. M., Langford, C. F., Pandian, R. D., Porter, K. M., Prigmore, E. (2006). The DNA sequence and biological annotation of human chromosome 1. *Nature* 441(7091), 315–321.

Güldener, U., Münsterkötter, M., Kastenmüller, G., Strack, N., Helden, J., Lemer, C., Richelles, J., Wodak, S. J., García-Martínez, J., Pérez-Ortín, J. E., Michael, H., Kaps, A., Talla, E., Dujon, B., André, B., Souciet, J. L., de Montigny, J., Bon, E., Gaillardin, C., Mewes, H. W. (2005). CYGD: the comprehensive yeast genome database. *Nucl. Acids Res.* 33, D364–D368.

Gupta, V. K., Tuohy, M. G., Ayyachamy, M., O'Donovan, A., Turner, K. M. (2013). *Laboratory protocols in fungal biology: current methods in fungal biology*, Springer, New York.

Hunter, S., Apweiler, R., Attwood, T. K., Bairoch, A., Bateman, A., Binns, D., Bork, P., Das, U., Daugherty, L., Duquenne, L., Finn, R. D., Gough, J., Haft, D., Hulo, N., Kahn, D., Kelly, E., Laugraud, A., Letunic, I., Lonsdale, D., Lopez, R., Madera, M., Maslen, J., McAnulla, C., McDowall, J., Mistry, J., Mitchell, A., Mulder, N., Natale, D., Orengo, C., Quinn, A. F., Selengut, J. D., Sigrist, C. J. A., Thimma, M., Thomas, P. D., Valentin, F., Wilson, D., Wu, C. H., Yeats, C. (2009). InterPro: the integrative protein signature database. *Nucl. Acids Res.* 37, D211–D215.

Inglis, D. O., Arnaud, M. B., Binkley, J., Shah, P., Skrzypek, M. S., Wymore, F., Binkley, G., Miyasato, S. R., Simison, M., Sherlock, G. (2012). The Candida genome database incorporates multiple *Candida* species: multispecies search and analysis tools with curated gene and protein information for *Candida albicans* and *Candida glabrata*. *Nucl. Acids Res.* 40(D1), D667–D674.

Karsch-Mizrachi, I., Nakamura, Y., Cochrane, G. (2012). The international nucleotide sequence database collaboration. *Nucl. Acids Res.* 40, D33–D37.

Kjøller, R., Bruns, T. D. (2003). Rhizopogon spore bank communities within and among Californian pine forests. *Mycologia* 95, 603–613.

Kodama, Y., Shumway, M., Leinonen, R. (2012). The sequence read archive: explosive growth of sequencing data. *Nucl. Acids Res.* 40, D54–D56.

Kõljalg, U., Larsson, K. H., Abarenkov, K., Nilsson, R. H., Alexander, I. J., Eberhardt, U., Erland, S., Høiland, K., Kjøller, R., Larsson, E., Pennanen, T., Sen, R., Taylor, A. F., Tedersoo, L., Vrålstad, T., Ursing, B. M. (2005). UNITE: a database providing web-based methods for the molecular identification of ectomycorrhizal fungi. *New Phytologist* 166, 1063–1068.

Kõljalg, U., Nilsson, R. H., Abarenkov, K., Tedersoo, L., Taylor, A. F. S., Bahram, M., Bates, S. T., Bruns, D. T., Bengtsson-Palme, J., Callaghan, M. T., Douglas, B., Drenkhan, T., Eberhardt, U., Dueñas, M., Grebenc, T., Griffith, W. G., Hartmann, M., Kirk, P. M., Kohout, P., Larsson, E., Björn, D. L., Lücking, R., Martín, M. P., Matheny, P. B., Nguyen, N. H., Niskanen, T., Oja, J., Peay, K. G., Peintner, U., Peterson, M., Põldmaa, K., Saag, L., Saar, I., Schüßler, A., Scott, A. J., Senés, C., Smith, M. E., Suija, A., Taylor, D. L., Telleria, M. T., Weiss, M., K.-Larsson, H. (2013). Towards a unified paradigm for sequence-based identification of fungi. *Molecular Ecology* 22(21), 5271–5277.

Larsson, K.-H., Larsson, E., Kõljalg, U. (2004). High phylogenetic diversity among corticioid homobasidiomycetes. *Mycological Research* 108, 983–1002.

Lawler, K., Hammond-Kosack, K., Brazma, A., Coulson, R. M. R. (2013). Genomic clustering and co-regulation of transcriptional networks in the pathogenic fungus *Fusarium graminearum*. *BMC Systems Biology* 7, 52, doi:10.1186/1752–0509-7-52.

Lilleskov, A., Fahey, T. J., Horton, T. R., Lovett, G. M. (2002). Belowground ectomycorrhizal community change over a nitrogen deposition gradient in Alaska. *Ecology* 83, 104–115.

Liu, X., Tang, W. H., Zhao, X. M., Chen, L. (2010). A network approach to predict pathogenic genes for *Fusarium graminearum*. *PLoS One* 5(10), e13021, doi:10.1371/journal.pone.0013021.

Lum, G., Min, X. J. (2011). FunSecKB: the Fungal Secretome KnowledgeBase. *Database – The Journal of Biological Databases and Curation*, doi: 10.1093/database/bar001.

Marthey, S., Aguileta, G., Rodolphe, F., Gendrault, A., Giraud, T., Fournier, E., Lopez-Villavicencio, M., Gautier, A., Lebrun, M., Chiapello, H. (2008). FUNYBASE: a FUNgal phYlogenomic database. *BMC Bioinformatics* 9, 456.

Matsuzaki, M., Misumi, O., Shin, T., Maruyama, S., Takahara, M., Miyagishima, S. Y., Mori, T., Nishida, K., Yagisawa, F., Nishida, K., Yoshida, Y., Nishimura, Y., Nakao, S., Kobayashi, T., Momoyama, Y., Higashiyama, T., Minoda, A., Sano, M., Nomoto, H., Oishi, K., Hayashi, H., Ohta, F., Nishizaka, S., Haga, S., Miura, S., Morishita, T., Kabeya, Y., Terasawa, K., Suzuki, Y., Ishii, Y., Asakawa, S., Takano, H., Ohta, N., Kuroiwa, H., Tanaka, K., Shimizu, N., Sugano, S., Sato, N., Nozaki, H., Ogasawara, N., Kohara, Y., Kuroiwa, T. (2004). Genome sequence of the ultrasmall unicellular red alga *Cyanidioschyzon merolae* 10D. *Nature* 428(6983), 653–657.

Meinken, J., Min, J. (2012). Computational prediction of protein subcellular locations in eukaryotes: an experience report. *Computational Molecular Biology* 2(1), 1–7.

Min, X. J. (2010). Evaluation of computational methods for secreted protein prediction in different eukaryotes. *J. Proteomics Bioinform.* 3, 143–147.

Moktali, V., Park, J., Fedorova-Abrams, N. D., Park, B., Choi, J., Y.-Lee, H., Kang, S. (2012). Fungal cytochrome P450 database 1.2. *BMC Genomics* 13, 525.

Moncalvo, J.-M., Vilgalys, R., Redhead, S. A., Johnson, J. E., James, T. Y., Aime, M. C., Hofstetter, V., Verduin, S. J. W., Larsson, E., Baroni, T. J., Thorn, R. G., Jacobsson, S., Clémençon, H., Miller, O. K. (2002). One hundred and seventeen clades of euagarics. *Molecular Phylogenetics and Evolution* 23, 357–400.

Nilsson, R. H., Larsson, K. H., Ursing, B. M. (2004). Galaxie-CGI scripts for sequence identification through automated phylogenetic analysis. *Bioinformatics* 20, 1447–1452.

Otávio J. B. B., Lucian, G. F., Cruz, D. C., Passos, F. M. L. (2009). Computational analysis of the interaction between transcription factors and the predicted secreted proteome of the yeast *Kluyveromyces lactis. BMC Bioinformatics* 10, 194, doi:10.1186/1471–2105–10–194.

Park, J., Park, J., Jang, S., Kim, S., Kong, S., Choi, J., Ahn, K., Kim, J., Lee, S., Kim, S., Park, B., Jung, K., Kim, S., Kang, S., Lee, Y. H. (2008a). FTFD: an informatics pipeline supporting phylogenomic analysis of fungal transcription factors. *Bioinformatics* 24(7), 1024–1025.

Park. J., Lee, S., Choi, J., Ahn, K., Park, B., Park, J., Kang, S., Lee, Y. H. (2008b). Fungal cytochrome P450 database. *BMC Genomics* 9, 402.

Parkinson, H., Sarkans, U., Kolesnikov, N., Abeygunawardena, N., Burdett, T., Dylag, M., Emam, I., Farne, A., Hastings, E., Holloway, E., Kurbatova, N., Lukk, M., Malone, J., Mani, R., Pilicheva, E., Rustici, G., Sharma, A., Williams, F., Adamusiak, T., Brandizi, M., Sklyar, N., Brazma, A. (2011). ArrayExpress update – an archive of microarray and high-throughput sequencing-based functional genomics experiments. *Nucl. Acids Res.* 39, D1002-D1004.

Parmasto, E., Nilsson, R. H., Larsson, K. H. (2004). Cortbase version 2, extensive updates of a nomenclatural database for corticioid fungi (Hymenomycetes). *Phyloinformatics* 1, 5.

Pearson, W. R. (2000). Flexible sequence similarity searching with the FASTA3 program package. *Methods Mol. Biol.* 132, 185–219.

Poisson, G., Chauve, C., Chen, X., Bergeron, A. (2007). FragAnchor: a large-scale predictor of glycozylphosphatidylinositol anchors in eukaryote protein sequences by qualitative scoring. *Genomics, Add, Proteomics and Bioinformatics* 5(2), 121–130.

Pruitt, K. D., Maglott, D. R. (2001). RefSeq and LocusLink: NCBI gene-centered resources. *Nucl. Acids Res.* 29(1), 137–140.

Pruitt, K. D., Tatusova, T., Brown, G. R., Maglott, D. R. (2012). NCBI Reference Sequences (RefSeq): current status, new features and genome annotation policy. *Nucl. Acids Res.* 40, D130-D135.

Pukkila, P. J., Skrzynia, C. (1993). Frequent changes in the number of reiterated ribosomal RNA genes throughout the life cycle of the basidiomycete *Coprinus cinereus. Genetics* 133, 203–211.

Quaedvlieg, W., Groenewald, J. Z., Yáñez-Morales, M. J., Crous, P. W. (2012). DNA barcoding of *Mycosphaerella* species of quarantine importance to Europe. *Persoonia* 29, 101–115.

Rastogi, G., Sani, R. K. (2011). Molecular techniques to assess microbial community structure, function, and dynamics in the environment. *In Microbes and microbial technology: agricultural and environmental applications*, ed. by Ahmad, I., Ahmad, F., Pichtel, J., Springer, New York, pp. 29–57.

Remm, M., Storm, C. E., Sonnhammer, E. L. (2001). Automatic clustering of orthologs and in-paralogs from pairwise species comparisons. *J. Mol. Biol.* 314, 1041–1052.

Robbertse, B., Tatusova, T. (2011). Fungal genome resources at NCBI. *Mycology* 2(3), 142–160.

Rose, P. W., Beran, B., Bi, C., Bluhm, W. F., Dimitropoulos, D., Goodsell, D. S., Prlic, A., Quesada, M., Quinn, G. B., Westbrook, J. D., Young, J., Yukich, B., Zardecki, C., Berman, H. M., Bourne, P. E. (2010). The RCSB Protein Data Bank: redesigned web site and web services. *Nucl. Acids Res.* 39, D392-D401.

Rosling, A., Landeweert, R., Lindahl, B. D., Larsson, K. H., Kuyper, T. W., Taylor, A. F. S., Finlay, R. D. (2003). Vertical distribution of ectomycorrhizal fungal taxa in a podzol soil profile. *New Phytologist* 159, 775–783.

Ruepp, A., Zollner, A., Maier, D., Albermann, K., Hani, J., Mokrejs, M., Tetko, I., Güldener, U., Mannhaupt, G., Münsterkötter, M., Mewes, H. W. (2004). The FunCat, a functional annotation scheme for systematic classification of proteins from whole genomes. *Nucl. Acids Res.* 32, 5539–5545.

Sangeetha, J., Thangadurai, D. (2013a). Staining techniques and biochemical methods for the identification of fungi. *In Laboratory protocols in fungal biology: current methods in fungal biology*, ed. by Gupta, V. K., Tuohy, M. G., Ayyachamy, M., Turner, K. M., O'Donovan, A., Springer, New York, pp. 237–258.

Sangeetha, J., Thangadurai, D. (2013b). Identification key for the major growth forms of lichenized fungi. *In Laboratory protocols in fungal biology: current methods in fungal biology*, ed. by Gupta, V. K., Tuohy, M. G., Ayyachamy, M., Turner, K. M., O'Donovan, A., Springer, New York, pp. 91–112.

Slippers, B., Boissin, E., Phillips, A. J. L., Groenewald, J. Z., Lombard, L., Wingfield, M. J., Postma, A., Burgess, T., Crous, P. W. (2013). Phylogenetic lineages in the Botryosphaeriales: a systematic and evolutionary framework. *Studies in Mycology* 76, 31–49.

Sonnenberg, A. S. M., Groot, P. W. J., Schaap, P. J., Baars, J. J. P., Visser, J., Griensven, L. J. L. D. (1996). Isolation of expressed sequence tags of *Agaricus bisporus* and their assignment to chromosomes. *Applied and Environmental Microbiology* 62, 4542–4547.

Srivastava, S. K., Huang, X., Brar, H. K., Fakhoury, A. M., Bluhm, H. B., Bhattacharyya, M. K. (2014). The genome sequence of the fungal pathogen *Fusarium virguliforme* that causes sudden death syndrome in Soybean. *PLoS One* 9(1), e81832, doi:10.1371/journal.pone.0081832.

Stajich, J. E., Harris, T., Brunk, B. P., Brestelli, J., Fischer, S., Harb, O. S., Kissinger, J. C., Li, W., Nayak, V., Pinney, D. F., Stoeckert, C. J., Roos, D. S. (2012). FungiDB: an integrated functional genomics database for fungi. *Nucl. Acids Res.* 40(D1), D675–D681.

Tanabe, A. S., Toju, H. (2013). Two new computational methods for universal DNA barcoding: a benchmark using barcode sequences of bacteria, archaea, animals, fungi, and land plants. *PLoS One* 8(10), e76910. doi:10.1371/journal.pone.0076910.

Tatusova, T., Ciufo, S., Fedorov, B., O'Neill, K., Tolstoy, I. (2014). RefSeq microbial genomes database: new representation and annotation strategy. *Nucl. Acids Res.* 42(1), D553–D559.

Tautz, D., Arctander, P., Minelli, A., Thomas, R. H., Vogler, A. P. (2003). A plea for DNA taxonomy. *Trends in Ecology and Evolution* 18, 70–74.

Tedersoo, L., Kõljalg, U., Hallenberg, N., Larsson, K. H. (2003). Fine scale distribution of ectomycorrhizal fungi and roots across substrate layers including coarse woody debris in a mixed forest. *New Phytologist* 159, 153–165.

Thompson, J. D., Higgins, D. G., Gibson, T. J. (1994). ClustalW: improving the sensitivity of progressive multiple sequence alignment through sequence weighting, position-specific gap penalties and weight matrix choice. *Nucl. Acids Res.* 22, 4673–4680.

Trail, F. (2009). For blighted waves of grain: *Fusarium graminearum* in the postgenomics era. *Plant Physiol.* 149(1), 103–110.

Tuskan, G. A., Difazio, S., Jansson, S., Bohlmann, J., Grigoriev, I., Hellsten, U., Putnam, N., Ralph, S., Rombauts, S., Salamov, A., Schein, J., Sterck, L., Aerts, A., Bhalerao, R. R., Bhalerao, R. P., Blaudez, D., Boerjan, W., Brun, A., Brunner, A., Busov, V., Campbell, M., Carlson, J., Chalot, M., Chapman, J., Chen, G. L., Cooper, D., Coutinho, P. M., Couturier, J., Covert, S., Cronk, Q., Cunningham, R., Davis, J., Degroeve, S., Déjardin, A., Depamphilis, C., Detter, J., Dirks, B., Dubchak, I., Duplessis, S., Ehlting, J., Ellis, B., Gendler, K., Goodstein, D., Gribskov, M., Grimwood, J., Groover, A., Gunter, L., Hamberger, B., Heinze, B., Helariutta, Y., Henrissat, B., Holligan, D., Holt, R., Huang, W., Islam-Faridi, N., Jones, S., Jones-Rhoades, M., Jorgensen, R., Joshi, C., Kangasjärvi, J., Karlsson, J., Kelleher, C., Kirkpatrick, R., Kirst, M., Kohler, A., Kalluri, U., Larimer, F., Leebens-Mack, J., Leplé, J. C., Locascio, P., Lou, Y., Lucas, S., Martin, F., Montanini, B., Napoli, C., Nelson, D. R., Nelson, C., Nieminen, K., Nilsson, O., Pereda, V., Peter, G., Philippe, R., Pilate, G., Poliakov, A., Razumovskaya, J., Richardson, P., Rinaldi, C., Ritland, K., Rouzé, P., Ryaboy, D., Schmutz, J., Schrader, J., Segerman, B., Shin, H., Siddiqui, A., Sterky, F., Terry, A., Tsai, C. J., Uberbacher, F., Unneberg, P., Vahala, J., Wall, K., Wessler, S., Yang, G., Yin, T., Douglas, C., Marra, M., Sandberg, G., Peer, Y., Rokhsar, D. (2006). The genome of black cottonwood, *Populus trichocarpa* (Torr. & Gray). *Science* 313(5793), 1596–1604.

van de Vossenberg, B. T. L. H., Westenberg, M., Bonants, P. J. M. (2013). DNA barcoding as an identification tool for selected EU-regulated plant pests: an international collaborative test performance study among 14 laboratories. *EPPO Bulletin* 43(2), 216–228.

Vrålstad, T., Myhre, E., Schumacher, T. (2002). Molecular diversity and phylogenetic affinities of symbiotic root-associated ascomycetes of the *Helotiales* in burnt and metal polluted habitats. *New Phytologist* 155, 131–148.

Will, K. W., Rubinoff, D. (2004). Myth of the molecule: DNA barcodes for species cannot replace morphology for identification and classification. *Cladistics* 20, 47–55.

Wong, P., Walter, M., Lee, W., Mannhaupt, G., Münsterkötter, M., Mewes, H. W., Adam, G., Güldener, U. (2011). FGDB: revisiting the genome annotation of the plant pathogen *Fusarium graminearum. Nucl. Acids Res.* 39(S1), D637–D639.

Wu, C. H., Apweiler, R., Bairoch, A., Natale, D. A., Barker, W. C., Boeckmann, B., Ferro, S., Gasteiger, E., Huang, H., Lopez, R., Magrane, M., Martin, M. J., Mazumder, R., O'Donovan, C., Redaschi, N., Suzek, B. (2006). The Universal Protein Resource (UniProt): an expanding universe of protein information. *Nucl. Acids Res.* 34, D187–D191.

Zdobnov, E. M., Apweiler, R. (2001). InterProScan – an integration platform for the signature-recognition methods in InterPro. *Bioinformatics* 17, 847–848.

CHAPTER 8

FUNGAL BIOMOLECULES AND FUNGAL SECONDARY METABOLITES

CONTENTS

8.1   INTRODUCTION

The various impacts of fungal molecules can be categorized into different levels such as degradation of pollutants, biofuel production, biopesticides and biocontrol agents, biofertilizers and secondary metabolites. Mycoremediation, a kind of bioremediation in which the fungi is used to degrade pollutants from the environment. It can be done by mixing mycelium into contaminated soil and placing mycelia mats over toxic sites. Fungi remove the heavy metals by channeling them into their fruit bodies and other pollutants by the enzymatic degradation. Among white-rot fungi and brown-rot fungi, white rot fungi are found to be more efficient in mycoremediation, which contain enzymes like lignin peroxidases, manganese peroxidases and laccases. A number of reviews dealing with degradation of environmental pollutants by white rot fungi have been published (Ponting, 2001). Extra-cellular lignin modifying enzyme produced by fungi shows the ability for degradation of all pollutants with structural similarity with lignin, because of its low substrate specificity. Thus fungi have an innate ability to degrade different recalcitrant pollutants. Moreover, their hyphal extension can reach pollutants in the soil thereby participate in ecological restoration.

8.2 FUNGAL BIORESOURCES

Spent mushroom compost has been utilized as the primary substrate in the treatment of coalmine drainage in constructed wetlands (Stark et al., 1994; Manyin et al., 1997), as an electron donor for the biological treatment of AMD and in the removal of heavy metals in passive treatments using laboratory columns and laboratory passive systems (Groudev et al., 1999). Spent mushroom compost is much more biodiverse and thus more effective than a pure culture in remediating a polluted site. The removal of organic toxic chemicals like di-n-butyl phthalate (DBP), di-2-ethyl hexyl phthalate (DEHP), nonylphenol (NP), and bisphenol-A (BPA) by laccase obtained from the spent mushroom compost (SMC) of the white rot fungi, *Ganoderma lucidum* were reported (Liao et al., 2012). The spent compost of oyster mushroom *Pleurotus pulmonarius* which was a degraded paddy straw-based substrate, contained 25% chitin could remove Pentachlorophenol (PCP), a widely used wood preservative since 1980s from PCP-contaminated water (Law et al., 2003).

Increasing energy cost, energy security and global warming concerns demand substitutes for petroleum-biofuel. Biofuel production from cellulosic material uses available substrates without competing with food supplies and therefore, presents an economic and environmental opportunity (Solomon, 2010). Recently, saprophytic zygomycete strain *Mucor indicus* has been identified as an ethanol-producing organism, capable to grow aerobically or anaerobically on a number of different carbon sources including hexoses and pentoses with yield and productivity in the same order as *Saccharomyces cerevisiae* (Sues et al., 2005). Lipids produced from filamentous fungi show great promise for biofuel production, but, a major limiting factor is the high production cost attributed to feedstock (Zheng et al., 2012). However, the cost will be reduced potentially if cheap feedstocks or waste materials can be used (Xue et al., 2008). By using a common fungus *Trichoderma* and a common bacterium *E. coli* together, isobutanol, a biofuel could be produced using stalks and leaves from corn plants as the raw material. In a study by Vicente et al. (2009), the filamentous fungus *Mucor circinelloides* was found to be a potential feedstock for biodiesel production. These microbial lipids showed a high content (>85%) of saponifiable matter and a suitable fatty acid profile for biodiesel production.

Insects and pests affecting crops can be controlled by fungi, such as Chinese caterpillar fungi by spraying spores of the fungi on the crop pests. Plant pathogen control also can be brought by *Trichoderma* like fungi against soil fungi like *Fusarium* and *Rhizoctonia*, using their hydrolytic enzymes thereby saving the plant from these pathogens effect. The antibiotic produced in them also act on other pathogens. Hence, plant pathogen control is possible by direct hyphal parasitism, antibiotics and competition like phenomenon showed by the fungal molecules. Use of *Trichoderma* in potting is found to be common now. This method is comparatively cheaper and less detrimental to the environment than using chemical pesticides. Plant parasitic nematodes can be controlled by fungi like *Psecilomyces lilacinus*. The toxins produced by different insect pathogens will establish the death of the host. Ascomyte fungi like *Beauvaria, Metarrhizium* and *Tolypocladium* are commonly used. Though the knockdown period is more and incomplete, the effect of the fungal biocontrol is more than the chemical application. By the biological control measures using fungi in turn help in reducing the application of chemicals for pest and pathogen control. Hence, it is considered as an important aspect for environmental protection by pollution control.

The use of chemical fertilizers and pesticides has caused tremendous harm to the environment. Environment friendly fertilizer is one of the alternatives for this. Fungi serve as a source for the enrichment of the soil and reduce the requirement of fertilizers in the soil. The most striking relationship that these have with plants is symbiosis, in which the partners derive benefits from each other. The contributions of fungi to the soil are by involving in nutrient cycle and also by decomposing the matters. Symbiotic relationship of mycorrhizal fungi with plant root can waive the amount of phosphate fertilizers, which is lacking in the soil of temperate region depleting and making available to the plants. It was found to be difficult in culturing Mycorrhizae. Normally nonsterile medium like soil is being used as the medium in the research. Due to its slow growth response, large volumes/weights of the inoculum have to be applied to the crop. Some of the fungi will retain in the soil to form humus and the remaining go into air where they can be used up as raw material for food synthesis. By liberating carbon dioxide these fungi participate in maintaining the never-ending cycle of carbon in nature. The carbon dioxide is very important for green plants in the preparation of food materials by photosynthesis. The recycling process would be reduced in the absence of fungi, which leads to piling up of non-degraded matters.

Fungi also act as a foe to the ecosystem as spreaders of plant diseases, animal mycoses, mycotoxins and spoilage. On the other hand, it is beneficial as a producer of useful metabolite and agent of biological control. The use of rust fungi, *Puccinia chondrillina* on Skeleton weed, *Chondrilla juncea* of wheat and *Colletotrichum orbiculare* on Bathurst bur, *Xanthium spinosum* has been successfully implemented. Specificity to the host, mutagenic ability of the pathogens, virulence, pathogenicity, reproduction and lifecycle should function well in the system. Moreover, if proper monitoring is also done, the spray of spore vegetable oil suspension will be effective.

8.3   FUNGAL BIOACTIVE SUBSTANCES

Secondary metabolites are compounds produced by an organism that are not required for primary metabolic processes. Fungi produce an enormous array of secondary metabolites, some of which are important in industry. Many fungi express secondary metabolites that influence competitive outcomes. The systematic study of secondary metabolites began in 1922, under the leadership of Harold Raistrick, who eventually characterized more than 200 mold metabolites (Raistrick, 1950). Among different classes of secondary metabolites such as polyketides, non-ribosomal peptides, terpenes and indole alkaloids, polyketides are the most abundant fungal secondary metabolites. They include, the yellow *A. nidulans* spore pigment intermediate naphthopyrone (WA), the carcinogen aflatoxin and the commercially important cholesterol lowering compound lovastatin (Kennedy, 1999). Plant hormones like auxin, cytokinins, gibberellins and abscisic acid were found to be produced in many pathogenic and benign fungi. The gibberellins were first found in the tall, straggly growth causing rice pathogenic fungus *Gibberella fujikuroi*. Apart from this, some of the fungi are found to be inducing the host metabolism; thereby, the growth of the plant such as formation of lateral roots and slowing of the root tip elongation by arbuscular mycorrhizae. The secondary metabolites produced by the fungi are enabling the recycling of the nutrients and ecosystem management. The metabolites produced in the contaminated food – toxins such as sporodesmin – are also to be considered, since, its effects are higher in the consumer.

The presence and metabolism of plant growth regulators in fungi are emphasizing the role of fungal molecule in sustainable agriculture. Although, fungi are useful in agriculture as biocontrol agent, they also have destructive effect as pathogen, which can lead to high loss in the cultivation. The majority of phytopathogenic fungi belong to the Ascomycetes or to the Basidiomycetes. Apart from this, fungal degraded agriwaste can be used in paper and fabric industries there by enabling recycling which would also contribute to the ecosystem management and environmental protection. Cultivation of edible fungi, mushroom and production of active ingredients are the major application of fungi in agriculture. It is necessary to analyze the impacts of the fungal molecule on ecosystem management and sustainable agriculture to bring forward the obscure metabolic pathway as well as the functions of these molecules. This chapter discusses the role of fungal enzymes on degradation of pollutants, in biofuel production, fungal secondary metabolites as biopesticides and biocontrol agents, fungal biomolecules as biofertilizers and ecological significance of fungal secondary metabolites. The chapter is an overview of applicability of fungal molecules on the current alternative demanded zones such as biofuel, bioferlizer and biocontrol.

KEYWORDS

- **Aflatoxin**
- **Ascomycetes**
- **Biocontrol agents**
- **Biofertilizers**
- **Biofuel production**

- **Biopesticides**
- **Brown-rot fungi**
- **Chinese caterpillar fungi**
- ***Chondrilla juncea***
- ***Colletotrichum orbiculare***
- **Ecological restoration**
- **Ecosystem management**
- **Edible fungi**
- **Environmental protection**
- **Enzymatic degradation**
- **Filamentous fungi**
- **Fungal biomolecules**
- **Fungal enzymes**
- **Fungal secondary metabolites**
- ***Ganoderma lucidum***
- ***Gibberella fujikuroi***
- **Hydrolytic enzymes**
- **Hyphal extension**
- **Isobutanol**
- **Laccases**
- **Lignin peroxidases**
- **Manganese peroxidases**
- **Mucor circinelloides**
- ***Mucor indicus***
- **Mycoremediation**
- **Mycorrhizal fungi**
- **Mycotoxins**
- **Non-ribosomal peptides**
- **Nutrient cycle**
- **Pentachlorophenol**
- **Phytopathogenic fungi**
- **Plant growth regulators**

- *Pleurotus pulmonarius*
- **Polyketides**
- *Psecilomyces lilacinus*
- *Puccinia chondrillina*
- **Recalcitrant pollutants**
- *Saccharomyces cerevisiae*
- **Spent mushroom compost**
- **Sporodesmin**
- **Substrate specificity**
- **Sustainable agriculture**
- **Symbiotic relationship**
- **Trichoderma**
- **White-rot fungi**
- *Xanthium spinosum*

REFERENCES

Groudev, S. N., Bratcova, S. G., Komnitsas, K. (1999). Treatment of waters polluted with radioactive elements and heavy metals by means of a laboratory passive system. *Minerals Engineering* 12(3), 261–270.

Kennedy, J. (1999). Modulation of polyketide synthase activity by accessory proteins during lovastatin biosynthesis. *Science* 284, 1368–1372.

Law, W. M., Lau, W. N., Lo, K. L., Wai, L. M., Chiu, S. W. (2003). Removal of biocide pentachlorophenol in water system by the spent mushroom compost of *Pleurotus pulmonarius*. *Chemosphere* 52(9), 1531–1537.

Liao, C. S., Yuan, S. Y., Hung, B. H., Chang, B. V. (2012). Removal of organic toxic chemicals using the spent mushroom compost of *Ganoderma lucidum*. *J. Environ. Monit.* 14(7), 1983–1988.

Manyin, T., Williams, F. M., Stark, L. R. (1997). Effects of iron concentration and flow rate on treatment of coal mine drainage in wetland mesocosms: an experimental approach to sizing of constructed wetlands. *Ecological Engineering* 9(3), 171–185.

Pointing, S. B. (2001). Feasibility of bioremediation by white-rot fungi. *Applied Microbiology and Biotechnology* 57, 20–33.

Raistrick, H. (1950). A region of biosynthesis. *Proc. R. Soc. Lond. B. Biol. Sci.* 136, 481–508.

Redecker, D., Kodner, R., Graham, L. E. (2000). Glomalean fungi from the Ordovician. *Science* 289, 1920–1921.

Solomon, B. D. (2010). Biofuels and sustainability. *Ann. N.Y. Acad. I. Sci.* 1185, 119–134.

Stark, L. M., Wenerick, W. R., Williams, F. M., Stevens, S. E., Wuest, P. J. (1994). Restoring the capacity of spent mushroom compost to treat coal mine drainage by reducing the inflow rate: a microcosm experiment. *Water, Air and Soil Pollution* 75(3–4), 405–420.

Sues, A. (2005). Ethanol production from hexoses, pentoses, and dilute-acid hydrolyzate by *Mucor indicus*. *FEMS Yeast Research* 5, 669–676.

Vicente, G., Bautista, L. F., Rodríguez, R., Gutiérrez, F. J., Sádaba, I., Ruiz-Vázquez, R. M., Torres-Martínez, S., Garre, V. (2009). Biodiesel production from biomass of an oleaginous fungus. *Biochemical Engineering Journal* 48(1), 22–27.

Xue, F., Miao, J., Zhang, X., Luo, H., Tan, T. (2008). Studies on lipid production by *Rhodotorula glutinis* fermentation using monosodium glutamate wastewater as culture medium. *Bioresour. Technol.* 99, 5923–5927.

Zheng, Y., Yu, X., Zeng, J., Chen, S. (2012). Feasibility of filamentous fungi for biofuel production using hydrolysate from dilute sulfuric acid pretreatment of wheat straw. *Biotechnology for Biofuels* 5, 50.

CHAPTER 9

FUNGAL BIOMOLECULES FOR DEGRADATION OF XENOBIOTICS AND ECOSYSTEM MANAGEMENT

CONTENTS

9.1   INTRODUCTION

In recent days, enzyme treatment opens up a new approach for the treatment of diverse environmental pollutants. Naturally, enzymes are most efficient tool for decomposition, soil remediation and breaking down toxic substances. In this purpose, fungi are being investigated for their capacity to degrade recalcitrant environmental pollutants such as aromatic hydrocarbons, hexogen, dyes and pesticides. Due to their ease of colonization, reproduction by numerous spores, and their metabolic versatility, fungi are ideally suited for bioremediation of environmental pollutants, which are difficult to treat by other means. Moreover, enzymes are both economically and environmentally beneficial because they are safely inactivated and create little or no waste; rather than being discarded, end-product enzymatic material may be treated and used as fertilizer. Enzyme research using fungi has been very active and promising in recent years (Maire et al., 2012). Crude enzyme of fungal strain *Fusarium* has been identified for the degradation of chlorpyrifos insecticide. Rate of degradation for chlorpyrifos by its intracellular enzyme, extracellular enzyme and cell fragment was calculated as 60.8%, 11.3% and 48%, respectively (Xie et al., 2005). Ligninolytic fungi, which are causing white rot disease in wood have been shown to degrade and mineralize versatile of environmental pollutants due to the non-specificity of their enzyme activity (Pointing, 2001). However, industrial applications of enzymes have delayed due to both fundamental and practical issues, such as enzyme stability and availability (Ayala et al., 2008). Fungal enzymes involved in the remediation of environmental pollutants are listed in Table 9.1.

TABLE 9.1 Fungal Enzymes Involved in biotransformation and Biodegradation of Environmental Pollutants

Fungus	Enzyme	Function	References
Agaricus bisporus	Polyphenol oxidase	Degradation of phenol	Singh (2006)
Aspergillus flavipes	Tyrosinase	Oxidation of phenol	Gukasyan (2002)
Aspergillus nidulans	N-Acetyl-6-hydroxytryptophan oxidase	Degradation of phenol	Birse and Clutterbuck (1990)
Bjerkandera adusta	Laccase, manganese peroxidase	Biotransformation of pesticides, decolorization of Reactive Blue	Heinfling-Weidtmann et al. (2001); Davila-Vazquez et al. (2005)
Caldariomyces fumago	Chloroperoxidase	Biohalogenation of phenol	Wannstedt et al. (1990)
Candida cylindracea	Esterase	Biotransformation of malathion	Singh (2006)
Ceriporiopsis subvermispora	Manganese peroxidase	Degradation of polycyclic aromatic hydrocarbons	Ruttimann-Johnson et al. (1994)
Coprinus cinereus	Peroxidase	Degradation of phenol	Budde et al. (2001)

TABLE 9.1 Continued

Fungus	Enzyme	Function	References
Coprinus macrorhizus	Peroxidase	Degradation of phenol	Al-Kassim et al. (1994)
Coriolopsis gallica	Laccase	Degradation of polycyclic aromatic hydrocarbons, decolorization of mixture of dyes	Pickard et al. (1999)
Coriolus hirsutus	Tyrosinase, laccase	Biosensor for determination of phenol, catechol and hydroquinone	Yaropolov et al. (1995)
Coriolus versicolor	Laccase	Degradation of trichlorophenol	Leontievsky et al. (2000)
Fusarium oxysporum	Cutinase	Biotransformation of malathion	Kim et al. (2005)
Ganoderma valesiacum	Manganese peroxidase, laccase	Degradation of polycyclic aromatic hydrocarbons	Nerud et al. (1991)
Lentinula edodes	Manganese peroxidase, laccase, β-glucosidase	Degradation of pentacholorophenol	Makkar et al. (2001)
Neurospora crassa	Polyphenol oxidase	Degradation of phenol	Singh (2006)
Panus tigrinus	Manganese peroxidase	Degradation of trichlorophenol	Leontievsky et al. (2000)
Penicillium simplicissimum	Methyltransferase, vanilyl-alcohol oxidase	Degradation of phenol	Jong et al. (1992)
Phanerochaete chrysosporium	Laccase, peroxidase	Biotransformation of pesticides, decolorization of dyes, degradation of pentacholorophenol	Cripps et al. (1990); Lin et al. (1991); Mougin et al. (2000)
Phlebia brevispora	Manganese peroxidase	Degradation of polycyclic aromatic hydrocarbons	Ruttimann et al. (1992)
Pleurotus eryngii	Laccase, peroxidase	Degradation of phenol and polycyclic aromatic hydrocarbons	Munoz et al. (1997); Rodriguez et al. (2004)

TABLE 9.1 Continued

Fungus	Enzyme	Function	References
Pleurotus ostreatus	Manganese peroxidase, lignin peroxidase	Biotransformation of biphenol A, decolorization of Congo Red and Methyl Orange	Shin and Kim (1998); Hirano et al. (2000)
Pycnosporus cinnabarinus	Laccase	Degradation of polycyclic aromatic hydrocarbons	Rama et al. (1998)
Rigidoporus lignosus	Manganese peroxidase, laccase	Degradation of polycyclic aromatic hydrocarbons	Galliano et al. (1991)
Stereum hirsutum	Manganese peroxidase	Degradation of polyaromatic hydrocarbons	Nerud et al. (1991)
Trametes versicolor	Laccase	Biotransformation of pesticides, degradation of pentachlorophenol and polycyclic aromatic hydrocarbons	Konishi and Inoue (1972); Jolivalt et al. (2000)

9.2 FUNGAL BIOMOLECULES AND DEGRADATION OF XENOBIOTICS

Peroxidases are enzymes that utilize hydrogen peroxide or other peroxides to catalyze the free-radical mediated oxidation of a variety of organic and inorganic compounds. Various types of fungal peroxidases differ based on the nature of their substrates. Fungal peroxidases are generally used in environmental cleanup, which is having important potential in transforming xenobiotics and other polluting compounds (Ayala et al., 2008). Since 1902, TNT (2,4,6-trinitrotoluene) has been used as an explosive. Normally, it is present in a crystalline form in soil due to its low water solubility. TNT is toxic to all life-forms and thus considered as a serious environmental hazard. Mineralization of TNT is achieved only by the activity of lignin peroxidase (LiP) and manganese peroxidases (MnP) of white-rot Basidiomycota. However, since these fungi are particularly sensitive to TNT and possess only a very limited ability to compete in the soil environment, TNT mineralization is unlikely to become a feasible tool for bioremediation in its own right. The most efficient TNT biotransformers belong to a species of *Absidia, Cunninghamella, Mortierella* (Zygomycota) and *Acremonium, Cylindrocarpon, Gliocladium* and *Trichoderma* (Ascomycota). Many fungi capable of biotransforming TNT are also capable of mineralizing hexogen. It is likely that a nitrate reductase is involved in both processes, in different ways, based on the stability of the ring system in these two substances (Weber et al., 2002).

Lignin in wood which are cross-linked to each other by a variety of different chemical bonds as a complex polymer of phenylpropane units. It is particularly persistent to degradation, and also reduces the bioavailability of the other cell wall constituents. Some higher fungi have enzymes, which are suitable for a selective chemical degradation of wood (Tuor et al., 1995). The fungal enzymes digest the brown lignin in wood, leaving the white cellulose behind for use in making paper. Wood rotting fungi are the only known organisms capable of degrading lignin by producing lignin-modifying enzymes, such as laccases, peroxidases and H_2O_2-generating oxidases. These enzymes can be used for biopulping, biobleaching, biotransformation and bioremediation. LiP catalyzes one-electron oxidations of phenolic and non-phenolic compounds. Typical reactions catalyzed by lignin peroxidases are ca-cb cleavage, ca oxidation, alkyl aryl cleavage, aromatic ring cleavage, demethylation, hydroxylation and polymerization (Chung and Aust, 1995; Singh, 2006). Lignolytic fungi such as *Phanerochaete chrysosporium, Coriolopsis polyzona, Pleurotus ostreatus* and *Trametes versicolor* typically secrete one or more of the three principal ligninolytic enzymes (Hatakka, 1994), such as lignin peroxidase (LiP), Mn-dependent peroxidase (MnP) and phenol oxidase (laccase) (LAC) (Thurston, 1994; Orth and Tien, 1995). These enzymes are involved in *in vitro* transformation of nitrotoluenes, PAHs, organic and synthetic dyes and pentachlorophenol (Lin et al., 1990; Hammel et al., 1991; Ollikka et al., 1993; Johannes et al., 1996; Heinfling et al., 1998; Acken et al., 1999; Novotny, 2004; Singh, 2006).

MnP have lower redox potentials than LiP and catalyze the Mn-mediated oxidation of lignin and phenolic compounds. Chloroperoxidase is secreted by the filamentous fungus *Caldariomyces fumago*. It catalyzes oxidative chlorination and in the absence of Cl⁻, enantioselective oxygen transfer reactions (Anke et al., 2003). Also LiP and MnP of *Phanerochaete chrysosporium* capable of decolorization of olive mill wastewater (Sayadi and Ellouz, 1995). Annibale et al. (2006) has found highest LiP activity in the *Phlebia* sp. and MnP activity in *Stachybotrys* sp. strain DABAC3. These two enzymes are actively involved in the biotransformation of pesticides (Singh, 2006) and degradation of aromatic compounds (Paszczynski et al., 1986). Commonly, peroxidases such as LiP and MnP are involved in the process of decolorization of dyes. These enzymes can cleave the aromatic rings and have the potential to remove color from the dyes. Azure B, Tropaeolin O, Methelene Blues, Orange II and sulfonated azo dyes are partially decolorized within 20 minutes by crude LiP produced by *Phanerochate chrysosporium* (Cripps et al., 1990). LiP from *Trametes versicolor* are decolorizing Remazol Brilliant Blue R (Christian et al., 2005), also *Bjerkandera adusta* shows degradation capability of Reactive Blue 38 and Reactive Violet 5 (Heinfling et al., 1998).

Laccase belong to the group of enzymes called blue copper oxidases. They are mostly extracellular glycoproteins with molecular weights between 60 and 80 kDa. Laccase catalyze four one-electron oxidations with redox potentials up to 0.8 V. Artificial substrates such as ABTS (2,2′-azino-bis-3-ethylbenzthiazoline-6-sulphonic acid) can act as mediators enabling the oxidation of non-phenolic compounds which cannot be oxidized by laccase on their own, thereby expanding the range of applications of these enzymes. Main application of white rot fungi and their oxidative enzymes is in biobleaching and biopulping in the pulp and paper industry, where they can replace environmentally harmful chemicals (e.g., chlorine) as well as saving energy costs of mechanical pulping. The enzyme laccase produced from different fungi was used to make paper. This process led to 30% reduction in energy consumption, 50% reduction in chemical product usage and a greater resistance for tearing (Maire et al., 2012). These are also useful for the degradation of many persistent organic pollutants. Commonly, fungal strain of *Pycnosporus cinnabarinus* is well known for the laccase degradation of polychlorinated biphenyls (Jonas et al., 2000; Schultz et al., 2001; Singh, 2006).

Laccase from *Pycnosporus cinnabarinus*, is used in the transformation of chlorinated hydroxybiphenyls to oligomerization products. Laccase is well documented as a mediator for oxidation process of certain aromatic compounds. White rot-fungus, *Trametes hirsuta* produces laccase, which is very well employed in the oxidation of alkenes (Niku-Paavola and Viikari, 2000). Extracellular laccase from *Phanerochaete chrysosporium* and *Trametes versicolor* are capable of degradation of diketonitrile in pesticide, thus, acting as a redox mediator (Mougin et al., 2000; Singh, 2006).

In lipase biotechnology, bioremediation of waste is a new and upcoming development. Lipase from different origin can be used in the oil spills resulting from rigging and refining, oil-wet night soils and shore sand, and lipid-tinged wastes in lipid processing factories and restaurants can be treated by the use of lipases of different origins (Sarada and Joseph, 1993). The increasing use of lipases in bioremediation has achieved greater importance with its successful application in the upgrading of waste fat (Salleh et al., 1993). Lipases are commonly produced by *Aspergillus niger*, *Aspergillus fumigatus* and *Mucor geophillus* (Naqvi et al., 2012). Phenol oxidases from *Pleurotus ostreatus* has been involved in the treatment of olive mill waste (OMW) (Martirani et al., 1996). Incubation of OMW with phenol oxidase catalyzed in an undetectable transformation (Singh, 2006). Several strains of *Trichosporon* yeast has been involved in the environmental degradation as biosensors (Neujahr, 1990). *Trichosporon cutaneum* produces phenol oxidases, which is used in the detection of phenol in the environment (Skladal, 1991; Canofeni et al., 1994). Polyphenol oxidase (PPO) or tyrosinase is a copper enzyme widely distributed in fungi such as common mushroom *Agaricus bisporus* and the bread mold *Neurospora crassa*. PPO catalyzes the hydroxylation of monophenols in the environment (Singh, 2006).

9.3 FUNGAL SECONDARY METABOLITES AND ECOSYSTEM MANAGEMENT

In most cases, fungi are under heavy attack by fungal grazers such as micro-, meso-, and macrofaunal elements, including protozoa and nematodes, mites and collembola, and earthworms and insects, respectively in soil ecosystems (Fox and Howlett, 2008; Stadler and Keller, 2008; Kempken and Rohlfs, 2010; Trienens et al., 2010). Additionally, saprotrophic filamentous fungi exploiting rich food sources like fruits, seeds and carrion may engage in competitive interactions with animals that depend on the same resources (Janzen, 1977; Rozen et al., 2008; Kempken and Rohlfs, 2010; Rohlfs and Churchill, 2011). These can seriously harm fungi in different phylogenetic affiliations and negatively affect fungal evolutionary fitness (Guevara et al., 2000; Rohlfs, 2005; Rohlfs et al., 2005; Tordoff et al., 2006; Gonigle, 2007; Boddy and Jones, 2008; Kempken and Rohlfs, 2010). Moreover, fungi share diverse evolutionary relationships with animals in general and arthropods in particular through induced response and volatile signaling, where fungi serve as a food source to fungal grazers, compete with saprophagous insects, and attack insect hosts for growth and reproduction (Rohlfs and Churchill, 2011). Some non-pathogenic fungi have taken living arthropods as a hub for growth and reproduction (Roy et al., 2006; Vega, 2008), while some pathogenic fungi are highly specialized natural enemies of arthropods (Veen et al., 2008; Samish, 1999), and others have been able to exploit both dead and living resources associated with plants and soils for resource exploitation (Leger, 2008; Rohlfs and Churchill, 2011).

Since endophytic fungi have the ability to alter feeding behavior of invertebrates and reduce invertebrate growth rates, fungal secondary metabolites play a key vital role in maintaining plant-fungus symbiosis, and thus protecting plants from herbivory (Schardl, 1996; Lane et al., 2000; Rodriguez et al., 2009; Rohlfs and Churchill, 2011). Although, we often consider fungal secondary metabolites as antifungal, antibacterial, antiviral, cytotoxic and immunosuppressive agents (Demain, 1999; Firáková et al., 2007; Ramos and Said, 2011), secondary metabolites of fungi influence the environment on a large scale as part of the nutrient cycle in ecosystems (Barea et al., 2005; Gadd, 2007; Lindahl et al., 2007). They help to control the population of damaging pests as animal pathogens and are specific to the insects they attack, and do not infect animals or plants (Tanada and Kaya, 1993; Sexton and Howlett, 2006). Fungal secondary metabolites aid in solving problems of ecosystems and environment (Barea et al., 2005; Gadd, 2007). In recent years, fungal secondary metabolites are predicted to yield a deeper understanding of the life histories of plant-fungal interactions and their potential effect on soil invertebrates (Lussenhop, 1992; Maraun et al., 2003; Thomas et al., 2012). Analogous to their hypothesized benefits in plants, secondary metabolites generated by fungi have an enormous impact on ecosystem management by serving as a chemical shield that fends off fungal feeders or competing saprophagous animals (Lindahl et al., 2007; Kempken and Rohlfs, 2010). Many secondary metabolites of fungi are of great ecological importance; they are decomposers in most ecosystems and are essential for the growth of most plants (Barea et al., 2005; Lindahl et al., 2007). A few fungal secondary metabolites stimulate sporulation and influence the development of producing organism, in addition to neighboring members of the same species; thereby enhance the fitness of a community of related species.

In boreal forest soils, fungal secondary metabolites play key role in organic matter decomposition, nutrient uptake, nutrient transfer and cycling of organic and inorganic nutrients, biogenic mineral formation, as well as, transformation and accumulation of metals. These mineral horizons of boreal forests and associated mycorrhizal mycelia transfer protons and organic metabolites from plant photosynthates to mineral surfaces, resulting in mineral dissolution, mobilization and redistribution of anionic nutrients and metal cations. In addition, mycorrhizal mycelia provide efficient systems for the uptake and direct transport of mobilized essential nutrients to their host plants (Finlay et al., 2009). The significant role of fungal secondary metabolites in resistance against competing insects are directly or indirectly related to mycotoxin synthesis, which has been well documented using an insect-fungus competition model system of *Aspergillus nidulans* and *Drosophila melanogaster* larvae. There are several ecologically important areas that benefit from application of the fungal secondary metabolites, particularly, biological control, sustainable forestry and land reclamation (Dodd and Thomson, 1994). For instance, fungi either directly or indirectly interact with all other organisms in an ecosystem in order to regulate several ecosystem processes. Fungi and secondary metabolites thereof, are crucial in the process of nutrient cycling, mineralization and immobilization of other elemental constituents as decomposers. They can also directly influence species composition and population dynamics of other organisms with which they coexist as parasites, pathogens, predators, mutualists and food sources. Moreover, they may act as agents of successional change or as factors contributing to ecosystem stability. In their natural environment, for example *Aspergillus* sp., produce aflatoxin as they face more competition from bacteria in alkaline soils and therefore, provide protection against insects in acidic environments. This would allow the fungus to establish itself in the community through protection against other eukaryotic competitors (Calvo et al., 2002).

Biosynthesis of these underexplored fungal secondary metabolites is favored by natural selection that increases its fitness under challenging ecological conditions, where in, antagonistic interactions with other organisms co-occurring in the fungal habitat have strong impact on their evolutionary fitness (Kempken and Rohlfs, 2010). Although, the specific functional and physiological role of secondary metabolites in the fungi that produce them remains mystery for many decades; the advantage is that they allow the organism to best survive in its ecological niche. Interestingly, recent studies have produced results that fungal secondary metabolites are evolved to protect fungi in such harsh environments with diverse array of competing organisms such as, amoebae, nematodes or invertebrates that can feed on fungi (Fox and Howlett, 2008). Secondary metabolites produced by the endophytic fungi such as, *Epichloë* and *Neotyphodium* spp., found in fescue grasses greatly reduce associated herbivorous insect populations, and thereby, increase plant fitness while reducing insect fitness (Gallery et al., 2007; Dalling et al., 2011). In an another non-grass system, researchers have reported the anti-feeding activity of rugulosin, a toxin produced by *Phialocephala scopiformis* that inhabits the white spruce needles against forest pest spruce budworm (*Choristoneura fumiferana*), suggesting the potential role of secondary metabolites of fungi in protection of forests against insect pests (Kempken and Rohlfs, 2010). Interestingly, secondary metabolites limit the burden of and have a negative effect on the animal antagonists in saprophytic, facultative or obligatory entomopathogenic fungi. The secondary metabolites of *Metarhizium anisopliae* and *Beauveria bassiana* are the two best-known examples of putative resistance mechanisms against animal antagonists.

9.4 FUNGAL ENZYMES IN BIOFUEL PRODUCTION

Lignocelluloses are one of the most abundant carbohydrate sources in plants and has significantly involved in conversion into liquid or biofuels. Biofuels are providing solution to reduce global emissions of greenhouse gases into the environment due to the usage of fossil fuels (Lynd et al., 1991; Demain et al., 2005; Yeoman et al., 2010). Plant polysaccharides have applications in many industrial sectors, such as biofuel, pulp and paper, food and feed. Enzymatic conversion of these polysaccharides in to lignocellulosic biomass will be a key technology in future biorefineries. Naturally, many fungal species playing vital role in the degradation of plant biomass (Sørensen et al., 2013).

Filamentous fungi are attractive resource for new enzymes due to its capability to grow on a wide range of substrates and efficiently degrade biopolymers. The decomposition of cellulosic plant biomass to glucose monomers for biofuel production is a typical example for an application that requires an enzyme-based approach in order to specifically cleave the glycosidic bonds between the glucose monomers of the cellulose chain and release single glucose molecules. The main enzymes necessary to degrade cellulosic plant material are cellulases and hemicellulases (Harman and Kubicek, 1998; Bouws et al., 2008; Kumar et al., 2008; Seiboth et al., 2011).

Fungal species are producing an extensive set of enzymes called carbohydrate-active enzymes dedicated to degrade specifically plant polysaccharides (Brick and Vries, 2011). Enzymatic degradation has advantages over chemical hydrolysis, as enzymes target specific linkages of the pectin molecules, while chemical methods are less specific (Schols and Voragen, 1996; Benen et al., 2002; Benoit et al., 2012). Direct conversion of biochemically-stored energy from renewable biomass resources into electricity is carried out by enzymatic biofuel cells (BFCs) (Bullen et al., 2006). However, enzyme purification is time-consuming and expensive process for this purpose. Furthermore, enzyme degradation occurred due to the long-term use of enzymatic BFCs, which limits their lifetime to only a few weeks. Recent research shows that crude culture supernatant from enzyme-secreting *Trametes versicolor* can be used without further treatment to supply the enzyme laccase to the cathode of a mediator less BFC (Jolivalt et al., 2005). The possibility to establish simple, cost efficient, and mediator less enzymatic BFC cathodes that do not require expensive enzyme purification procedures also can be achieved (Sane et al., 2013).

Bioethanol made from lignocellulosic biomass is considered to be the most promising biofuels. However, one of the major barriers for the production of biofuel economically is the enzymatic hydrolysis of the cellulose component to liberate glucose for ethanol fermentation because of the recalcitrance of feedstock (Wang et al., 2012). Lignocellulosic materials are the essential feedstock for second-generation biofuels. One of the current limitations to the production process for second-generation biofuels is linked to the poor performance of the enzymatic cocktails used to break down lignocellulosic biomass into fermentable monosaccharides. Catalases are enzymes that catalyze the depolymerization of cellulose. However, complete and efficient hydrolysis of cellulose is assisted by three cellulolytic enzyme activities, namely endoglucanase (1,4-β-D-glucan glucohydrolase [EC 3.2.1.4]), exoglucanase (1,4-β-D-glucan cellobiohydrolase [EC 3.2.1.91]), and β-glucosidase (β-D-glucoside glucohydrolase, [EC3.2.1.21]). Exoglucanase is produced by many fungi, such as, *Thermoascus aurantiacus, Talaromyces emersonii* and *Cladosporium* sp. Endoglucanase is produced mainly by thermophillic fungi such as *Caldocellulosiruptor saccharolyticus, Chaetomium thermophilum, Syncephalastrum racemosum, Talaromyces emersonii, Thermoascus aurantiacus* and *Thermomonospora curvata*. Glucosidases are produced mainly by thermophillic fungi, such as, *Thermoascus aurantiacus, Talaromyces emersonii, Sclerotium rolfsii, Paecilomyces thermophila* and *Monascus purpureus*. Fungal β-glucosidases play a vital role in the hydrolysis of cellulosic biomass for producing the monomer sugars for the production of biofuels and also provide platform molecules that can serve as building blocks in the synthesis of chemicals and polymeric materials (Himmel et al., 2007; Yeoman et al., 2010).

Hemicellulose, a highly branched mixture of complex polysaccharides such as, xylans, glucans, xyloglucans, callose, mannans and glucomannans. Optimizing the enzymatic conversion of lignocelluloses to fermentable sugars must take into an account of hemicellulose depolymerization (Chesson et al., 1986; Marcus et al., 2008). Xylanases have long been utilized in industries like food, paper, and fine chemical production. Further, they act as a critical component in the deconstruction of lignocellulose for biofuel production (Garcia-Aparicio et al., 2007). Fungal xylanases can be used in the production of biofuel due to its stronger catalytic activities (Lee et al., 2009). The xylanases from *Aspergillus awamori, Bispora* sp., and *Neurospora crassa* exhibit specific catalytic activity (Kormelink et al., 1993). The *Bispora* sp. xylanase, Xyn10C, in particular, appears to be an attractive option for biotechnological adaptation. This enzyme displays optimal activity at 80°C, higher than any other fungal xylanase, and is active over a broad pH range (pH 1.5–6.0) (Luo et al., 2009; Yeoman et al., 2010).

A thermophilic β-xylosidase from the fungus *Scytalidium thermophilum* was, however, found to be immune to xylose-mediated conversion (Zanoelo et al., 2004). This character is likely to be immensely important in the biofuel industry to the overall efficiency of biocatalysts. Thermostable α-L-arabinofuranosidases have also been found in *Penicillium capsulatum* exhibiting optimal activity at 60°C and 55°C (Filho et al., 1996). In addition, a novel thermostable α-L-arabinofuranosidase from *Aspergillus pullulans* can hydrolyze arabinan and debranched arabinan and was shown to have optimal activity at 75°C. Genus *Trichoderma* fungus, which produces enzymes capable of breaking down the cellulose and xylan chains into sugar molecules, has been utilized in biofuel production. However, the fungus does not always produce these enzymes; production must be stimulated using an inductor (disaccharide sophorose) (Saha and Bothast, 1998; Yeoman et al., 2010).

An enzyme blend to transform tough, woody plant material such as corn stalks and wood chips into fuel is considered a key component in the commercialization of second-generation biofuels. *Trichoderma reesei* produces enzymes that can breakdown lignocelluloses, the tough structural material that makes up plant cell walls, while *Aspergillus* species, produce many enzymes to degrade pectin (Martens-Uzunova and Schaap, 2009; Brink and Vries, 2011; Peterson and Nevalainen, 2012) as well as *Rhizopus* sp. mainly degrades the homogalacturonan part of pectin (Battaglia et al., 2011; Benoit et al., 2012). Lignocelluloses need to be broken down in order to get fermentable sugars that can be made into biofuels. The fungus produces dozens of these enzymes, each of which can attack the lignocelluloses differently. Recently, chemists are trying to combine and improve upon the best enzymes to create a chemical cocktail that could be used in biofuel production in a way that is simple and cost effective (Martinez et al., 2008; Kubicek et al., 2011).

One of the biggest hurdles to achieve global fuel need, lies in optimizing the multistep process involved in breaking down plant biomass and then converting it. Many of the cellulases currently used in biofuel production are derived from species that thrive at temperatures of 20–35°C (68–95°F), which is a room temperature to nearly body temperature. The conversion process at these temperatures takes time, during which contaminants can reduce the final yield. To speed up the conversion process, temperature has to be increased, which in turn, requires enzymes that are stable above current working conditions. Genomes of *Thielavia terrestris* and *Myceliophthora thermophila* were tested to thrive in high-temperature environments above 45°C and whose enzymes, active at temperatures ranging from 40 to 75°C, would therefore be useful for accelerating (thus improving) the biofuel production process. Thermostable enzymes and thermophilic cell factories may afford economic advantages in the production of many chemicals and biomass-based fuels. The 38.7-million base pair (Mbp) genome of thermophilic *Myceliophthora thermophila* and the 36.9 Mbp genome of *Thielavia terrestris* encodes multitude of enzymes that decompose biomass material. These two thermophiles can be considered all-purpose decomposers with respect to their carbohydrate-active enzymes (CAZymes) and their ability to degrade plant polysaccharides (Gilbert, 2011).

The enzyme cellulase from *Myceliophthora thermophila* and *Thielavia terrestris* have evolved efficiently in breaking down and in converting biomass into simple sugars at a wide range of temperatures. Since, these thermophiles are much more efficient than other cellulose degraders in breaking down cellulosic biomass, their enzymes are likely to be more active than known cellulases and they have developed new strategies for biomass degradation. These thermophilic fungi represent excellent hosts for biorefineries, where biomass is converted to biofuels as an alternative to modern oil refineries. These fungi are classified as rare organisms known as thermophiles, which thrive at temperatures between 45°C and 122°C. A key component of these organisms ability to survive is the fact that their enzymes can still function even under temperatures considered extreme. These enzymes are known as cellulases and are the ones currently used in biofuel production thrive at temperatures of 20–35°C. The low temperature threshold of cellulase is considered unfortunate as studies have shown that, a high-temperature environment is better for biofuel production (Berka et al., 2011). New fungal enzymes, notably β-xylosidases, glycoside hydrolases and oxidases from *Aspergillus japonicus* and *Trichoderma reesei* are being used for the production of biofuels (Semenova et al., 2009).

Biofuel cells produce electricity that is environmentally friendly and conserve resources, for instance, from organic waste material. They can use enzymes as catalysts to enable electrochemical reactions that generate electricity. In contrast to precious metal catalysts in conventional fuel cells, these enzymes can be obtained at low cost from renewable raw materials. For many technical applications, however, their lifetime is too short. The new concept developed to solve this problem by ensuring that the fuel cell is continually resupplied with the biocatalyst. Biofuel cell consists of laccase produced by *Trametes versicolor*-based cathode and glucose oxidase produced by *Aspergillus niger* based anode. These fungi release the enzymes into a solution surrounding the cathode where it enables the electrochemical conversion of oxygen. By comparison, the cathodes only have a lifetime of 14 days if they are not supplied with more enzymes. Since, the enzymatic solution can be supplied directly to the fuel cell without time-consuming and expensive purification, the costs are reduced to a minimum (Barriere et al., 2006). Enzyme production in biorefineries is achieved by culturing fungi in different media such as, *Trichoderma reesei* cultured on pretreated wheat straw (Gyalai-Korpos et al., 2011), *Aspergillus niger* and *Aspergillus saccharolyticus* (Sørensen et al., 2011) cultured on the fiber waste fraction left after hydrolysis and fermentation, and *Aspergillus japonicus* cultured on castor bean meal waste for biodiesel production (Herculano, 2011).

KEYWORDS

- *Agaricus bisporus*
- *Alaromyces emersonii*
- **Animal antagonists**
- **Anionic nutrients**
- **Anti-feeding activity**
- **Aromatic hydrocarbons**
- *Aspergillus awamori*
- *Aspergillus fumigatus*

- *Aspergillus japonicas*
- *Aspergillus nidulans*
- *Aspergillus niger*
- *Aspergillus pullulans*
- *Aspergillus saccharolyticus*
- **Biobleaching**
- **Bioethanol**
- **Biofuel**
- **Biological control**
- **Biopolymers**
- **Biopulping**
- **Biorefineries**
- **Bioremediation**
- **Biotransformation**
- *Bjerkandera adusta*
- **Blue copper oxidases**
- **Boreal forest soils**
- **Brown lignin**
- *Caldariomyces fumago*
- *Caldocellulosiruptor saccharolyticus*
- **Callose**
- **Carbohydrate-active enzymes**
- **Cellulases**
- **Cellulose degraders**
- *Chaetomium thermophilum*
- *Choristoneura fumiferana*
- *Coriolopsis polyzona*
- **Decomposers**
- **Depolymerization**
- *Drosophila melanogaster*
- **Ecosystem processes**
- **Ecosystem stability**

- Mannans
- *Metarhizium anisopliae*
- Mineral dissolution
- Mn-dependent peroxidase
- *Monascus purpureus*
- *Mucor geophillus*
- *Myceliophthora thermophila*
- Mycorrhizal mycelia
- Mycotoxin synthesis
- *Neurospora crassa*
- Nutrient cycling
- Oil refineries
- Olive mill waste
- Organic metabolites
- Oxidative chlorination
- *Paecilomyces thermophila*
- Pathogenic fungi
- *Penicillium capsulatum*
- Peroxidases
- Pesticides
- *Phanerochaete chrysosporium*
- Phenol oxidase
- *Phialocephala scopiformis*
- Phylogenetic affiliations
- Plant biomass
- Plant polysaccharides
- Plant-fungus symbiosis
- *Pleurotus ostreatus*
- Polychlorinated biphenyls
- Population dynamics
- *Pycnosporus cinnabarinus*

REFERENCES

Acken, B., Godefroid, L. M., Peres, C. M., Naveau, H., Agathos, S. N. (1999). Mineral-ization of 14C-U ring labeled 4-hydroxylamino-2,6-dinitrotoluene by manganese-dependent peroxidase of the white-rot Basidiomycete, *Phlebia radiata. J. Biotechnol.* 68, 159–169.

Anke, H., Kuhn, A., Weber, R. W. S. (2003). The role of nitrate reductase in the degrada-tion of the explosive RDX (hexahydro-1,3,5-trinitro-1,3,5-triazine) by *Penicillium* sp. AK96151. *Mycological Progress* 2, 219–225.

Annibale, A., Rosetto, F., Leonardi, V., Federici, F., Petruccioli, M. (2006). Role of autoch-thonous filamentous fungi in bioremediation of a soil. *Applied and Environmental Microbiology* 72(1), 28–36.

Ayala, M., Pickard, M. A., Vazquez-Duhalt, R. (2008). Fungal enzymes for environmental purposes, a molecular biology challenge. *J. Mol. Microbiol. Biotechnol.* 15(2–3), 172–180.

Barea, J. M., Pozo, M. J., Azcón, R., Azcón-Aguilar, C. (2005). Microbial co-operation in the rhizosphere. *Journal of Experimental Botany* 56(417), 1761–1778.

Barrière, F., Kavanagh, P., Leech, D. (2006). A laccase-glucose oxidase biofuel cell prototype operating in a physiological buffer. *Electrochimica Acta* 51(24), 5187–5192.

Battaglia, E., Benoit, I., Brink, J., Wiebenga, A., Coutinho, P. M., Henrissat, B., Vries, R. P. (2011). Carbohydrate-active enzymes from the zygomycete fungus *Rhizopus oryzae*: a highly specialized approach to carbohydrate degradation depicted at genome level. *BMC Genomics* 12, 38.

Benen, J. A. E., Vincken, J. P., Alebeek, G., J. W. M. (2002). Microbial pectinases, *In Pec-tins and their manipulation*, ed. by Seymour, G. B., Knox, J. P., Blackwell Publishing Ltd., Oxford, pp. 174–221.

Benoit, I., Coutinho, P. M., Schols, H. A., Gerlach, J. P., Henrissat, B., Vries, R. P. (2012). Degradation of different pectins by fungi: correlations and contrasts between the pectinolytic enzyme sets identified in genomes and the growth on pectins of different origin. *BMC Genomics* 13, 321–331.

Berka, R. M., Grigoriev, I. V., Otillar, R., Salamov, A., Grimwood, J., Reid, I., Ishmael, N., John, T., Darmond, C., Moisan, M. C., Henrissat, B., Coutinho, P. M., Lombard, V., Natvig, D. O., Lindquist, E., Schmutz, J., Lucas, S., Harris, P., Powlowski, J., Bellemare, A., Taylor, D., Butler, G., Vries, R. P., Allijn, I. E., Brink, J. (2011). Com-parative genomic analysis of the thermophilic biomass-degrading fungi *Mycelioph-thora thermophila* and *Thielavia terrestris. Nature Biotechnology* 29, 922–927.

Boddy, L., Jones, T. H. (2008). Interactions between basidiomycota and invertebrate. *In Ecology of Saprotrophic Basidiomycetes*, ed. by Boddy, L., Frankland, J. C., West, P., Elsevier, Amsterdam, pp. 156–174.

Bouws, H., Wattenberg, A., Zorn, H. (2008). Fungal secretomes – nature's toolbox for white biotechnology. *Applied Microbiology and Biotechnology* 80(3), 381–388.

Brink, J., Vries, R. P. (2011). Fungal enzyme sets for plant polysaccharide degradation. *Appl. Microbiol. Biotechnol.* 91, 1477–1492.

Bullen, R. A., Arnot, T. C., Lakeman, J. B., Walsh, F. C. (2006). Biofuel cells and their development. *Biosens. Bioelectron.* 21(11), 2015–2045.

Calvo, A. M., Wilson, R. A., Bok, J. W., Keller, N. P. (2002). Relationship between secondary metabolism and fungal development. *Microbiol Mol Biol Rev.* 66(3), 447–459.

Canofeni, S., Di, S. S., Mela, J., Pilloton, R. (1994). Comparison of immobilization procedures of development of an electrochemical PPO-based biosensor for online monitoring of a depuration process. *Anal. Lett.* 27, 1659–1669.

Chesson, A., Stewart, C. S., Dalgarno, K., King, T. P. (1986). Degradation of isolated grass mesophyll, epidermis and fiber cell-walls in the rumen and by cellulolytic rumen bacteria in axenic culture. *J. Appl. Bacteriol.* 60, 327–336.

Christian, V., Shrivastava, R., Shukla, D., Modi, H., Vyas, B. R. M. (2005). Mediator role of veratryl alcohol in the lignin peroxidase-catalyzed oxidative decolorization of Remazol Brilliant Blue R. *Enzyme Microb. Technol.* 36, 426–431.

Chung, N., Aust, S. D. (1995). Veratryl alcohol-mediated indirect oxidation of phenol by lignin peroxidases. *Arch. Biochem. Biophys.* 316, 733–737.

Cripps, C., Bumpus, J. A., Aust, S. D. (1990). Biodegradation of azo and heterocyclic dyes by *Phanerochaete chrysosporium. Appl. Environ. Microbiol.* 56, 1114–1118.

Dalling, J. W., Davis, A. S., Schutte, B. J., Arnold, A. E. (2011). Seed survival in soil: interacting effects of predation, dormancy and the soil microbial community. *Journal of Ecology* 99(1), 89–95.

Demain, A. L. (1999). Pharmaceutically active secondary metabolites of microorganisms. *Applied Microbiology and Biotechnology* 52, 455–463.

Demain, A. L., Newcomb, M., Wu, J. H. D. (2005). Cellulase, clostridia and ethanol. *Microbiol. Mol. Biol. Rev.* 69, 124–154.

Dodd, J. C., Thomson, B. D. (1994). The screening and selection of inoculant arbuscular-mycorrhizal and ectomycorrhizal fungi. *Plant and Soil* 159(1), 149–158.

Filho, E. X., Puls, J., Coughlan, M. P. (1996). Purification and characterization of two arabinofuranosidases from solid-state cultures of the fungus *Penicillium capsulatum. Appl. Environ. Microbiol.* 62, 168–173.

Finlay, R., Wallander, H., Smits, M., Holmstrom, S., Hees, P., Lian, B., Rosling, A. (2009). The role of fungi in biogenic weathering in boreal forest soils. *Fungal Biology Reviews* 23(4), 101–106.

Firáková, S., Sturdíková, M., Múcková, M. (2007). Bioactive secondary metabolites produced by micro-organisms associated with plants. *Biologia (Bratislava)* 62, 251–257.

Fox, E. M., Howlett, B. J. (2008). Secondary metabolism: regulation and role in fungal biology. *Current Opinion in Microbiology* 11(6), 481–487.

Gadd, G. M. (2007). Geomycology: biogeochemical transformations of rocks, minerals, metals and radionuclides by fungi, bioweathering and bioremediation. *Mycological Research* 111(1), 3–49.

Gallery, R. E., Dalling, J. W., Arnold, A. E. (2007). Diversity, host affinity and distribution of seed-infecting fungi: a case-study with *Cecropia. Ecology* 83, 582–588.

Garcia-Aparicio, M. P., Ballesteros, M., Manzanares, P., Ballesteros, I., Gonzalez, A., Negro, M. J. (2007). Xylanase contribution to the efficiency of cellulose enzymatic hydrolysis of barley straw. *Appl. Biochem. Biotechnol.* 137–140(1–12), 353–365.

Gilbert, D. (2011). Tagging enzymes that can take the heat. *The Primer* 8(4), 1.

Guevara, R., Rayner, A. D. M., Reynolds, S. E. (2000). Effects of fungivory by two specialist ciid beetles (*Octotemnus glabriculus* and *Cis boleti*) on the reproductive fitness of their host fungus, *Coriolus versicolor. New Phytologist* 145, 137–144.

Gyalai-Korpos, M., Mangel, R., Alvira, P., Dienes, D., Ballesteros, M., Reczey, K. (2011). Cellulase production using different streams of wheat grain- and wheat straw-based ethanol processes. *J. Ind. Microbiol. Biotechnol.* 38, 791–802.

Hammel, K. E., Green, B., Gai, W. Z. (1991). Ring fission of anthracene by a eukaryote. *Proceedings of the National Academy of Sciences USA* 88, 10605–10608.

Harman, G. E., Kubicek, C. P. (1998). *Trichoderma* and *Gliocladium*: enzymes, biological control and commercial applications, Taylor and Francis Ltd, London, Great Britain.

Heinfling, A., Martinez, M. J., Martinez, A. T., Bergbauer, M., Szewzyk, U. (1998). Transformation of industrial dyes by manganese peroxidases from *Bjerkandera adusta* and *Pleurotus eryngii* in a manganese-independent reaction. *Appl. Environ. Microbiol.* 64, 2788–2793.

Herculano, P. N., Porto, T. S., Moreira, K. A., Pinto, G. A. S., Souza-Motta, C. M., Porto, A. L. F. (2011). Cellulase production by *Aspergillus japonicus* URM5620 using waste from castor bean (*Ricinus communis* L.) under solid-state fermentation. *Appl. Biochem. Biotechnol.* 165, 1057–1067.

Himmel, M. E., Ding, S. Y., Johnson, D. K., Adney, W. S., Nimlos, M. R., Brady, J. W., Foust, T. D. (2007). Biomass recalcitrance: engineering plants and enzymes for biofuel production. *Science* 315(5813), 804–807.

Janzen, D. (1977). Why fruits rot, seeds mold, and meat spoils. *The American Naturalist* 111, 691–713.

Jolivalt, C., Madzak, C., Brault, A., Caminade, E., Malosse, C., Mougin, C. (2005). Expression of laccase IIIb from the white-rot fungus *Trametes versicolor* in the yeast *Yarrowia lipolytica* for environmental applications. *Applied Microbiology and Biotechnology* 66, 450–456.

Jolivalt, C., Brenon, S., Caminade, E., Mougin, C., Pontie, M. (2000). Immobilization of laccase from *Trametes versicolor* on a modified PVDF microfiltration membrane: characterization of the grafted support and application in removing a phenylurea pesticide in wastewater. *J. Membr. Sci.* 180, 103–113.

Jonas, U., Hammer, E., Haupt, E. T. K., Schauer, F. (2000). Characterization of coupling products formed by biotransformation of biphenyl and diphenyl ether by the white rot fungus *Pycnosporus cinnabarinus*. *Arch. Microbiol.* 174, 393–398.

Kempken, F., Rohlfs, M. (2010). Fungal secondary metabolite biosynthesis – a chemical defense strategy against antagonistic animal? *Fungal Ecology* 3(3), 107–114.

Kormelink, F. J., Leeuwen, M. J. F. S., Wood, T. M., Voragen, A. G. J. (1993). Purification and characterization of three endo-1,4-β-xylanases and one β-xylosidase from *Aspergillus awamori*. *J. Biotechnol.* 27, 249–265.

Kubicek, C. P., Herrera-Estrella, A., Seidl-Seiboth, V., Martinez, D. A., Druzhinina, I. S., Thon, M., Zeilinger, S., Casas-Flores, S., Horwitz, B. A., Mukherjee, P. K., Mukherjee, M., Kredics, L., Alcaraz, L. D., Aerts, A., Antal, Z., Atanasova, L., Cervantes-Badillo, M. G., Challacombe, J., Chertkov, O., McCluskey, K., Coulpier, F., Deshpande, N., von Doehren, H., Ebbole, D. J., Esquivel-Naranjo, E. U., Fekete, E., Flipphi, M., Glaser, F., Gomez-Rodriguez, E. Y., Gruber, S., Han, C., Henrissat, B., Hermosa, R., Hernandez-Onate, M., Karaffa, L., Kosti, I., Le Crom, S., Lindquist, E., Lucas, S., Lubeck, M., Lubeck, P. S., Margeot, A., Metz, B., Misra, M., Nevalainen, H., Omann, M., Packer, N., Perrone, G., Uresti-Rivera, E. E., Salamov, A., Schmoll, M., Seiboth, B., Shapiro, H., Sukno, S., Tamayo-Ramos, J. A., Tisch, D., Wiest, A., Wilkinson, H. H., Zhang,

M., Coutinho, P. M., Kenerley, C. M., Monte, E., Baker, S. E., Grigoriev, I. V. (2011). Comparative genome sequence analysis underscores mycoparasitism as the ancestral life style of *Trichoderma*. *Genome Biology* 12(4), R40.

Kumar, R., Singh, S., Singh, O. V. (2008). Bioconversion of lignocellulosic biomass: biochemical and molecular perspectives. *J. Ind. Microbiol. Biotechnol.* 35(5), 377–391.

Lane, G. A., Christensen, M. J., Miles, C. O. (2000). Coevolution of fungal endophytes with grasses: the significance of secondary metabolites. *In Microbial Endophytes*, ed. by Bacon, C. W., White, J. F. J., Marcel Dekker, New York, pp. 341–388.

Lee, J. W., Park, J. Y., Kwon, M., Choi, I. G. (2009). Purification and characterization of a thermostable xylanase from the brown-rot fungus *Laetiporus sulfureus*. *J. Biosci. Bioeng.* 107, 33–37.

Leger, R. J. (2008). Studies on adaptations of *Metarhizium anisopliae* to life in the soil. *J. Invert. Pathol.* 98, 271–276.

Lin, J. E., Wang, H. Y., Hickey, R. F. (1990). Degradation kinetics of pentachlorophenol by *Phanerochaete chrysosporium*. *Biotechnol. Bioeng.* 35, 1125–1134.

Lindahl, B. D., Ihrmark, K., Boberg, J., Trumbore, S. E., Högberg, P., Stenlid, J., Finlay, R. D. (2007). Spatial separation of litter decomposition and mycorrhizal nitrogen uptake in a boreal forest. *New Phytologist* 173(3), 611–620.

Luo, H., Li, J., Yang, J., Wang, H., Yang, Y., Huang, H., Shi, P., Yuan, T., Fan, Y., Yao, B. (2009). A thermophilic and acid stable family-10 xylanase from the acidophilic fungus *Bispora* sp. MEY-1. *Extremophiles* 13, 849–857.

Lussenhop, J. (1992). Mechanisms of micro-arthropod-microbial interactions in soil. *Adv. Ecol. Res.* 23, 1–33.

Lynd, L. R., Cushman, J. H., Nichols, R. J., Wyman, C. E. (1991). Fuel ethanol from cellulosic biomass. *Science* 251, 1318–1323.

Maire, V., Alvarez, G., Colombet, J., Comby, A., Despinasse, R., Dubreucq, E., Joly, M., Lehours, A. C., Perrier, V., Shahzad, T., Fontaine, S. (2012). An unknown respiration pathway substantially contributes to soil CO_2 emissions. *Biogeosciences Discuss.* 9, 8663–8691.

Maraun, M., Martens, H., Migge, S., Theenhaus, A., Scheu, S. (2003). Adding to 'the enigma of soil animal diversity': fungal feeders and saprotrophic soil invertebrates prefer similar food substrates. *Eur. J. Soil Biol.* 39, 85–95.

Marcus, S. E., Verhertbruggen, Y., Herve, C., Ordaz-Ortiz, J. J., Farkas, V., Pedersen, H. L., Willats, W. G., Knox, J. P. (2008). Pectic homogalacturonan masks abundant sets of xyloglucan epitopes in plant cell walls. *BMC Plant Biol.* 8, 60.

Martens-Uzunova, E. S., Schaap, P. J. (2009). Assessment of the pectin degrading enzyme network of *Aspergillus niger* by functional genomics. *Fungal. Genet. Biol.* 46(1), 170–179.

Martinez, D., Berka, R. M., Henrissat, B., Saloheimo, M., Arvas, M., Baker, S. E., Chapman, J., Chertkov, O., Coutinho, P. M., Cullen, D., Danchin, E. G., Grigoriev, I. V., Harris, P., Jackson, M., Kubicek, C. P., Han, C. S., Ho, I., Larrondo, L. F., Leon, A. L., Magnuson, J. K., Merino, S., Misra, M., Nelson, B., Putnam, N., Robbertse, B., Salamov, A. A., Schmoll, M., Terry, A., Thayer, N., Westerholm-Parvinen, A., Schoch, C. L., Yao, J., Barabote, R., Nelson, M. A., Detter, C., Bruce, D., Kuske, C. R., Xie, G., Richardson, P., Rokhsar, D. S., Lucas, S. M., Rubin, E. M., Dunn-Coleman, N., Ward, M., Brettin,

T. S. (2008). Genome sequencing and analysis of the biomass-degrading fungus *Trichoderma reesei* (syn. *Hypocrea jecorina*). *Nat. Biotechnol.* 26(5), 553–560.

Martirani, L., Giardina, P., Marzullo, L., Sannia, G. (1996). Reduction of phenol content and toxicity in olive mill waste waters with the ligninolytic fungus *Pleurotus ostreatus*. *Water Research* 30, 1914–1918.

Mougin, C., Boyer, F. D., Caminade, E., Rama, R. (2000). Cleavage of the diketonitrile derivative of the herbicide isoxaflutole by extra cellular fungal oxidases. *J. Agric. Food Chem.* 48, 4529–4534.

Neujahr, H. Y. (1990). Yeasts in biodegradation and biodeterioration process. *Bioprocess. Technol.* 5, 321–348.

Niku-Paavola, M. L., Viikari, L. (2000). Enzymatic oxidation of alkenes. *J. Mol. Catal. B. Enzymatic* 10, 435–444.

Novotný, Č., Svobodová, K., Erbanová, P., Cajthaml, T., Kasinath, A., Lang, E., Šašek, V. (2004). Ligninolytic fungi in bioremediation: extracellular enzyme production and degradation rate. *Soil Biology and Biochemistry* 36, 1545–1551.

Ollikka, P., Alhonmäki, K., Leppänen, V., Glumoff, T., Raijola, T., Suominen, I. (1993). Decolorization of azo, triphenylmethane, heterocyclic, and polymeric dyes by lignin peroxidase isoenzymes from *Phanerochaete chrysosporium*. *Applied and Environmental Microbiology* 59, 4010–4016.

Orth, A. B., Tien, M. (1995). Biotechnology of lignin degradation, *In The Mycota. II. Genetics and Biotechnology*, ed. by Esser, K., Lemke, P. A., Springer, Berlin, pp. 287–302.

Paszczynski, A., Huynh, V. B., Crawford, R. L. (1986). Comparisons of ligninase-I and peroxidize-M2 from the white-rot fungus *Phanerochaete chrysosporium*. *Arch. Biochem. Biophys.* 244, 750–765.

Peterson, R., Nevalainen, H. (2012). *Trichoderma reesei* RUT-C30 – thirty years of strain improvement. *Microbiology* 158, 58–68.

Pointing, S. B. (2001). Feasibility of bioremediation by white-rot fungi. *Applied Microbiology and Biotechnology* 57, 20–33.

Ramos, H. P., Said, S. (2011). Modulation of biological activities produced by an endophytic fungus under different culture conditions. *Advances in Bioscience and Biotechnology* 2, 443–449.

Rodriguez, E., Nuero, O., Guillen, F., Martinez, A. T., Martinez, M. J. (2004). Degradation of phenolic and non-phenolic aromatic pollutants by four *Pleurotus* species: the role of laccase and versatile peroxidase. *Soil Biol. Biochem.* 36, 909–916.

Rohlfs, M. (2005). Density-dependent insect-mold interactions: effects on fungal growth and spore production. *Mycologia* 97, 996–1001.

Rohlfs, M., Churchill, A. C. L. (2011). Fungal secondary metabolites as modulators of interactions with insects and other arthropods. *Fungal Genetics and Biology* 48, 23–34.

Rohlfs, M., Obmann, B., Petersen, R. (2005). Competition with filamentous fungi and its implications for a gregarious life-style in insects living on ephemeral resources. *Ecological Entomology* 30, 556–563.

Roy, H. E., Steinkraus, D. C., Eilenberg, J., Hajek, A. E., Pell, J. K. (2006). Bizarre interactions and endgames: entomopathogenic fungi and their arthropod hosts. *Annual Review of Entomology* 51, 331–357.

Rozen, D. E., Engelmoer, D. J. P., Smiseth, P. T. (2008). Antimicrobial strategies in burying beetles breeding on carrion. *Proceedings of the National Academy of Sciences of USA* 105, 17890–17895.

Saha, B. C., Bothast, R. J. (1998). Purification and characterization of a novel thermostable a-L-arabinofuranosidase from a color-variant strain of *Aureobasidium pullulans*. *Appl. Environ. Microbiol.* 64, 216–220.

Salleh, A. B., Musani, R., Basri, M., Ampon, K., Yunus, W. M. Z., Razak, C. N. A. (1993). Extra- and intra- cellular lipases from thermophilic *Rhizopus oryzae* and factors affecting their production. *Can. J. Microbiol.* 39, 978–981.

Samish, M. (1999). Pathogens and predators of ticks and their potential in biological control. *Annu. Rev. Entomol.* 44, 159–182.

Sane, S., Jolivalt, C., Mittler, G., Nielsen, P. J., Rubenwolf, S., Zengerle, R., Kerzenmacher, S. (2013). Overcoming bottlenecks of enzymatic biofuel cell cathodes: crude fungal culture supernatant can help to extend lifetime and reduce cost. *ChemSusChem.* 6(7), 1209–1215.

Sarada, R., Joseph, R. (1993). Profile of hydrolytic activity on macromolecules of tomato process made during anaerobic digestion. *Enzyme Microbiol. Technol.* 15, 339–342.

Sayadi, S., Ellouz, R. (1995). Roles of lignin peroxidase and manganese peroxidase from *Phanerochaete chrysosporium* in the decolorization of olive mill wastewaters. *Appl. Environ. Microbiol.* 61, 1098–1103.

Schardl, C. L. (1996). Epichloë species: fungal symbionts of grasses. *Annu. Rev. Phytopathol.* 34, 109–130.

Schols, H. A., Voragen, A. G. J. (1996). Complex pectins: structure elucidation using enzymes. *In Progress in Biotechnology: Pectins and pectinases, Vol. 14*, ed. by Visser, J., Voragen, A. G. J., Elsevier Science, Amsterdam, Netherlands, pp. 3–19.

Schultz, A., Jonas, U., Hammer, E., Schauer, F. (2001). Dehalogenation of chlorinated hydroxybiphenyls by fungal laccase. *Appl. Environ. Microbiol.* 67, 4377–4381.

Seiboth, B., Ivanova, C., Seidl-Seiboth, V. (2011). *Trichoderma reesei*: a fungal enzyme producer for cellulosic biofuels. *In Biofuel production – recent developments and prospects*, ed. by Bernardes, M. A. S., InTech, Croatia, pp. 309–341.

Semenova, M. V., Drachevskaya, M. I., Sinitsyna, O. A., Gusakov, A. V., Sinitsyn, A. P. (2009). Isolation and properties of extracellular β-xylosidases from fungi *Aspergillus japonicus* and *Trichoderma reesei*. *Biochemistry* 74(9), 1002–1008.

Sexton, A. C., Howlett, B. J. (2006). Parallels in fungal pathogenesis on plant and animal hosts. *Eukaryot Cell* 5(12), 1941–1949.

Singh, H. (2006). *Mycoremediation: fungal bioremediation*. John Wiley and Sons, Inc., London.

Skladal, P. (1991). Mushroom tyrosinase-modified carbon paste electrode as an amperometric biosensor for phenols. *Collect. Czech. Chem. Commun.* 56, 1427–1433.

Sørensen, A., Lübeck, M., Lübeck, P. S., Ahring, B. K. (2013). Fungal beta-glucosidases: a bottleneck in industrial use of lignocellulosic materials. *Biomolecules* 3, 612–631.

Sørensen, A., Teller, P. J., Lübeck, P. S., Ahring, B. K. (2011). Onsite enzyme production during bioethanol production from biomass: screening for suitable fungal strains. *Appl. Biochem. Biotechnol.* 164, 1058–1070.

Stadler, M., Keller, N. P. (2008). Paradigm shifts in fungal secondary metabolite research. *Mycol. Res.* 112, 127–130.

Tanada, Y., Kaya, H. K. (1993). *Insect Pathology*, Academic Press, Inc., San Diego.

Thomas, W. C., Lynne, B., Hefin, J. T. (2012). Functional and ecological consequences of saprotrophic fungus-grazer interactions. *ISME J.* 6(11), 1992–2001.

Thurston, C. F. (1994). The structure and function of fungal laccases. *Microbiology* 140, 19–26.

Tordoff, G. M., Boddy, L., Jones, T. H. (2006). Grazing by *Folsomia candida* (Collembola) differentially affects mycelial morphology of the cord-forming basidiomycetes *Hypholoma fasciculare, Phanerochaete velutina* and *Resinicium bicolor. Mycological Research* 110, 335–345.

Trienens, M., Keller, N. P., Rohlfs, M. (2010). Fruit, flies and filamentous fungi-experimental analysis of animal-microbe competition using *Drosophila melanogaster* and *Aspergillus* mold as a model system. *Oikos* 1, 1–11.

Tuor, U., Winterhalterb, K., Fiechter, A. (1995). Enzymes of white-rot fungi involved in lignin degradation and ecological determinants for wood decay. *J. Biotechnol.* 41(1), 1–17.

Veen, F. J. F., Müller, C. B., Pell, J. K., Godfray, H. C. J. (2008). Food web structure of three guilds of natural enemies: predators, parasitoids and pathogens of aphids. *Journal of Animal Ecology* 77, 191–200.

Vega, F. E. (2008). Insect pathology and fungal endophytes. *Journal of Invertebrate Pathology* 98, 277–279.

Wang, M., Li, Z., Fang, X., Wang, L., Qu, Y. (2012). Cellulolytic enzyme production and enzymatic hydrolysis for second-generation bioethanol production. *Adv. Biochem. Eng. Biotechnol.* 128, 1–24.

Weber, R. W. S., Ridderbusch, D. C., Anke, H. (2002). 2,4,6-Trinitrotoluene (TNT) tolerance and biotransformation potential of microfungi isolated from TNT-contaminated soil. *Mycological Research* 106, 336–344.

Xie, H., Zhu, L. S., Wang, J., Wang, X. G., Liu, W., Qian, B., Wang, Q. (2005). Enzymatic degradation of organophosphorus insecticide chlorpyrifos by fungus WZ-I. *Huan Jing Ke Xue* 26(6), 164–168.

Yeoman, C. J., Han, Y., Dodd, D., Schroeder, C. M., Mackie, R. I., Cann, I. K. O. (2010). Thermostable enzymes as biocatalysts in the biofuel industry. *Advances in Applied Microbiology* 70, 1–55.

Zanoelo, F. F., L. Polizeli Md Mde, Terenzi, H. F., Jorge, J. A. (2004). Purification and biochemical properties of a thermostable xylose-tolerant β-D-xylosidase from *Scytalidium thermophilum. J. Ind. Microbiol. Biotechnol.* 31, 170–176.

CHAPTER 10

FUNGAL SECONDARY METABOLITES FOR BIOCONTROL AND SUSTAINABLE AGRICULTURE

CONTENTS

10.1 INTRODUCTION

Soil-borne pathogenic fungi are responsible for severe damage to many agricultural and horticultural crops worldwide (Rossman, 2009). Commonly, plant diseases are controlled by using chemical pesticides or fungicides. However, these chemicals are proven harmful for the environment and have created a path to the development of biological control agents (BCAs), which is a potential alternative to the chemical pesticides. The selective activity of numerous fungal biomolecules like secondary metabolites against specific infection stages without accompanying toxicity against plant pathogens indicate directions for the development of future natural product-derived fungicides which are more easily degradable in the environment and possess fewer non-target effects. Such substances are produced by many saprotrophic and endophytic fungi in pure culture (Thines et al., 2004). Entomopathogenic fungi were among the first organisms to be used for the biological control of pests. More than 700 species of fungi from 90 genera are pathogenic to insects. Some examples of fungi and its biomolecules acting as biocontrol agents are given in Table 10.1.

10.2   FUNGAL BIOCONTROL AGENTS

Beauveria bassiana (Bals.) Vuill. is an important entomopathogenic mito-sporic ascomycete, acting as a natural pathogen of insects and it has been developed as a microbial insecticide for use against many major arthropod pests in agriculture (Faria and Wraight, 2007). Among mycoinsecticides,

TABLE 10.1   Potential Fungal Biomolecules as Biocontrol Agents against Plant Pathogens

Fungi	Biomolecules	Target organisms	References
Aschersonia aleyrodis	Proteases, collagenases, chymolesterases	Whitefly	Charnley and Leger (1991); Liu et al. (2006)
Beauveria bassiana	Chitinases, proteases, collagenases, chymolesterases, endonucleases, bassianin, bassiacridin, beauvericin, bassianolide, beauverolides, tenellin and oosporein	Gram pod borer, termites, thrips, whiteflies, aphids, plant hoppers	Khachatourians (1991); Kim et al. (1999); Yokoyama et al. (2002); Nahar et al. (2003)
Beauveria brongniartii	Proteases, collagenases, chymolesterases, bassianin, bassiacridin, beauvericin, bassianolide, beauverolides, tenellin and oosporein	Cockchafer larvae	Khachatourians (1996); Daniella (2000)
Entomophthora coronate	Proteases, collagenases, chymolesterases	Ants, planthoppers	Charnley and Leger (1991)

TABLE 10.1   Continued

Fungi	Biomolecules	Target organisms	References
Lagenidium giganteum	Proteases, collagenases, chymolesterases	Mosquito larvae	Charnley and Leger (1991)
Metarhizium anisopliae	Chitin deacetylase, proteases, collagenases, chymolesterases, superoxide dismutase	Beetles, termites, spittlebugs, gram pod borer, plant hopper	Khachatourians (1996); Nahar et al. (2004); Khan et al. (2012)
Nomuraea rileyi	Chitinase, proteases, collagenases, chymolesterases	Gram pod borer, *Rhipicephalus microplus*, *Spodoptera litura*	Nahar et al. (2003); Sheng et al. (2006)
Tolypocladium inflatum	Chitinase, proteases, collagenases, chymolesterases	Bibionid flies, mosquito larvae, beetle larvae	Hodge et al. (1996)
Trichoderma harzianum	Chitinolytic enzymes, trichodermin, harzianum A	*Sclerotium rolfsii*, *Rhizoctonia solani, Botrytis cinerea*	Elad et al. (1983); Cruz et al. (1992); Malmierca et al. (2012)
Trichoderma virens	Gliotoxin	*Sclerotium rolfsii*, *Rhizoctonia solani, Pythium* sp.	Wilhite et al. (2001)
Verticillium lecanii	Chitinase, proteases, collagenases, chymolesterases	Potato aphid, *Sphaerotheca fuliginea*	Sheng et al. (2006)

about 33.9% is based on *Beauveria bassiana*, followed by *Metarhizium anisopliae* (33.9%), *Isaria fumosorosea* (5.8%) and *Beauveria brongniartii* (4.1%) (Faria and Wraight, 2007). The initial interaction in the pathogenesis is mediated by mechanical force, enzymatic processes and certain metabolic acids (Qazi and Khachatourians, 2005). The enzymes involved in pathogenesis of insects are generally grouped into proteases, peptidases, chitinases and lipases.

The genus *Trichoderma*, a mycoparasitic fungi, particularly, *Trichoderma harzianum* have been used against *Rhizoctonia solani*, *Sclerotium rolfsii* and *Botrytis cinerea* for crop protection. Their abilities are a combination of several mechanisms, including nutrient competition and direct mycoparasistism, which involves the production of antifungal metabolites and cell wall biodegrading enzymes. In Southeast and Midwest regions of Brazil, soil fungi, which are causing plant diseases are responsible for great losses in common bean (*Phaseolus vulgaris* L.) in irrigated areas (Almeida et al., 2007; Qualhato et al., 2013). Schirmbock et al. (1994) conducted a study and reported that *Trichoderma harzianum* producing chitinase, β-1,3-glucanase and protease, utilize glucose as a carbon source which was active against *Botrytis cinerea* by inhibiting spore germination as well as hyphal elongation of *Botrytis cinerea*. Also concluded that, the enzymes and the peptaibols were tested together, an antifungal synergistic interaction was observed and 50% effective dose values obtained were in the range of those determined in the culture supernatants. Therefore, the parallel formation and synergism of hydrolytic enzymes and antibiotics may have an important role in the antagonistic action of *Trichoderma harzianum* against fungal phytopathogens.

10.3   FUNGAL BIOFERTILIZERS AND FUNGAL BIOPESTICIDES

In commercial biofertilizers and biopesticides, *Trichoderma harzianum* strains, T22 and T39 are used as active BCA and are widely applied amongst field and greenhouse crops. Vinale et al. (2006) studied the three major compounds produced by *Trichoderma harzianum* strain T22, one is a new azaphilone that shows *in vitro* inhibition against *Rhizoctonia solani*, *Pythium ultimum* and *Gaeumannomyces graminis* var. *tritici*. In turn, filtrates from strain *Trichoderma harzianum* strain T39 were demonstrated to contain two compounds previously isolated from other *Trichoderma harzianum* strains and a new butenolide. Also, reported the isolation and characterization of the main secondary metabolites obtained from culture filtrates of two *Trichoderma harzianum* strains and their production during antagonistic interaction with the pathogen *Rhizoctonia solani*. Mycoparasitism is a mechanism of fungi to control phytopathogen under the control of enzymes. Many fungi produce and release lytic enzymes that can hydrolyze a wide variety of polymeric compounds including chitin, protein, cellulose, hemicellulose and DNA. Expression and secretion of these enzymes by different fungi can sometimes result in the suppression of plant pathogen activities directly. Harman (2000) reported the involvement of chitinase and β-1,3-glucanase in the *Trichoderma* sp. mediated biological control. Since, enzymes are the products of genes, production of desired enzymes is achieved by slight change in the structure of gene. Gupta et al. (1995) reported that, a strain of *Trichoderma*, deficient in the ability to produce endo chitinase has reduced ability to control *Botrytris cineria*, but shows increased ability to control *Rhizoctonia solani*.

An efficient BCA is one that produces sufficient quantities of biomolecules such as enzymes, antibiotics or other secondary metabolites in the vicinity of the plant pathogen (Chaube et al., 2003). Commonly, BCA are producing three types of antibiotics such as, nonpolar or volatile, polar or non-volatile and water-soluble. Among all of these, the non-polar antibiotics are more effectively used as they can act at the sites away from the site of production. Several BCAs are capable to produce several antibiotics, which can suppress one or more pathogens separately or with combination.

BCA can increase the plant growth by reducing the occurrence of disease incidence of crops at least during the early stages of the life cycle by the way of disease escape. The best example is the resistance of damping off of Solanaceous crops with the advancing age. Chaube et al. (2003) reported that, fungal BCAs are managing the plant diseases by promoting the growth of plants through increased solubilization of nutrients, increased nutrient uptake through enhanced root growth and sequestration of nutrients. Notably, *Aspergillus niger* strain AN-27 was reported to produce plant growth promoting compounds such as, 2-carboxy-methyl-3-hexyl-maleic anhydride and 2-methylene-3-hexyl-butanedioic acid (hexylitaconic acid), that were directly responsible for increased root and shoot length and biomass of crop plants. *Trichoderma* preparations have been reported to increase vigor and emergence of oat seedlings. Seeds coated with *Trichoderma viride* increased fresh and dry weight of shoot, root and nodules of broadbeans (Woo et al., 2006). Also, salicylic acid and non-expressor of pathogenesis-related genes 1 (NPR1) are key players in systemic acquired resistance of *Trichoderma harzianum*; when inoculated on to roots or on to leaves of grapes control diseases caused by *Botrytis cineria*.

Many classes of compounds are released by the *Trichoderma* sp., into the zone of interaction and induce resistance in plants. The first class is proteins with enzymatic or other activity. Fungal proteins such as xylanase, cellulases and swollenins are secreted by *Trichoderma* species (Martinez et al., 2001). *Trichoderma* endochitinase can also enhance defense, probably, through induction of plant defense related proteins. Saksirirat et al. (2009) evaluated the efficacy of *Trichoderma* strains by inducing resistance in tomato. It was found that, the activity of the enzymes in the leaves of tomato increased upto 14th day. This indicates that, *Trichoderma* was effective in inducing systemic resistance in tomato plant.

The fungi from which protein degrading enzymes proteases, collagenases, and chymoleastases have been identified and characterized are *Aschersonia aleyrodis, Beauveria bassiana, Beauveria brongniartii, Entoophthora coronata, Nomuraea rileyi, Metarhizium anisopliae* and *Verticillium lecanii* (Sheng et al., 2006). Small and Bidochka (2005) identified that *Metarhizium anisopliae* contain genes that encodes extracellular subtilisin-like proteinase (Pr1) for cuticle degradation. Kim et al. (1999) described the gene structure and expression of a novel *Beauveria bassiana* protease (bassianin I) which is 1137 bp and 379 amino acids long. Endo and exo-chitinases play an important role in the cleavage of N-Acetylglucosamine (NAGA) polymer of the insect cuticle into smaller units or monomers, that the extracellular chitinases are virulence determinant factors (Khachatourians, 1991). Chitinolytic enzymes (N-acetyl-β-D-glucosaminidases) and endochitinases are commonly produced by *Metarhizium anisopliae, Metarhizium flavoviride* and *Beauveria bassiana*. Since, lipids represent major chemical constituents of the insect cuticle, enzymes capable of hydrolyzing these compounds, such as phospholipases, could be expected to be involved in the cuticle disruption processes that occur during host invasion. Phospholipases are a heterogeneous group of enzymes that are able to hydrolyze one or more ester linkages in glycerophospholipids. Phospholipase B (PLB) was secreted by different fungal species such as *Candida albicans, Aspergillus fumigatus* and *Cryptococcus neoformans*.

KEYWORDS

- Antifungal metabolites
- *Aschersonia aleyrodis*
- *Aspergillus fumigatus*
- *Aspergillus niger*
- Bassianin I
- *Beauveria bassiana*
- *Beauveria brongniartii*
- Biofertilizers
- Biological control agents
- Biopesticides
- *Botrytis cinerea*
- *Candida albicans*
- Cell wall degrading enzymes
- Cellulases
- Chitinases
- Chymoleastases
- Collagenases
- *Cryptococcus neoformans*
- Entomopathogenic fungi
- *Entoophthora coronate*
- Fungal biomolecules
- Fungal proteins
- *Gaeumannomyces graminis* var. *tritici*
- Hexylitaconic acid
- Hydrolytic enzymes
- Hyphal elongation
- *Isaria fumosorosea*
- Lipases
- *Metarhizium anisopliae*
- *Metarhizium flavoviride*

- Mycoinsecticides
- Mycoparasitism
- N-Acetylglucosamine
- *Nomuraea rileyi*
- Peptidases
- *Phaseolus vulgaris*
- Phospholipases
- Plant diseases
- Proteases
- Protein degrading enzymes
- *Pythium ultimum*
- *Rhizoctonia solani*
- *Sclerotium rolfsii*
- Soil-borne pathogenic fungi
- Spore germination
- swollenins
- *Trichoderma harzianum*
- *Trichoderma viride*
- *Verticillium lecanii*
- Xylanase
- β-1,3-glucanase

REFERENCES

Almeida, F. B. R., Cerqueira, F. M., Silva, R. N., Ulhoa, C. J., Lima, A. L. (2007). Myco-parasitism studies of *Trichoderma harzianum* strains against *Rhizoctonia solani*: evaluation of coiling and hydrolytic enzyme production. *Biotechnology Letters* 29(8), 1189–1193.

Chaube, H. S., Mishra, D. S., Varshney, S., Singh, U. S. (2003). Biocontrol of plant pathogens by fungal antagonists: a historical background, present status and future prospects. *Annual Review of Plant Pathology* 2, 1–42.

Faria, M. R., Wraight, S. P. (2007). Mycoinsecticides and mycoacaricides: a comprehensive list with worldwide coverage and international classification of formulation types. *Biological Control* 43, 237–256.

Gupta, S., Arora, D. K., Shrivastava, A. K. (1995). Growth promotion of tomato plants by *Rhizobacteria* and imposition of energy stress on *Rhizoctonia solani. Soil Biology and Biochemistry* 27, 1051–1058.

Harman, G. E. (2000). Myths and dogmas of biocontrol: changes in the perceptions derived from research on *Trichoderma harzianum* T-22. *Plant Disease* 84, 377–393.

Khachatourians, G. G. (1991). Physiology and genetics of entomopathogenic fungi, *In Handbook of Mycology*, ed. by Arora D. K., Mukerji, K. G., Drouchet, E., Marcel Dekker, New York, pp. 613–663.

Kim, H. K., Hoe, H. S., Suh, D. S., Kang, S. C., Hwang, C., Kwon, S. T. (1999). Gene structure and expression of the gene from *Beauveria bassiana* encoding bassiasin I, an insect cuticle-degrading serine protease. *Biotechnol. Lett.* 21, 777–783.

Martinez, C., Blanc, F., Le, C. E., Besnard, O., Nicole, M., Baccou, J. C. (2001). Salicylic acid and ethylene pathways are differentially activated in melon cotyledons by active or heat-denatured cellulase from *Trichoderma longibrachiatum. Plant Physiology* 127, 334–344.

Martinez, D., Berka, R. M., Henrissat, B., Saloheimo, M., Arvas, M., Baker, S. E., Chapman, J., Chertkov, O., Coutinho, P. M., Cullen, D., Danchin, E. G., Grigoriev, I. V., Harris, P., Jackson, M., Kubicek, C. P., Han, C. S., Ho, I., Larrondo, L. F., Leon, A. L., Magnuson, J. K., Merino, S., Misra, M., Nelson, B., Putnam, N., Robbertse, B., Salamov, A. A., Schmoll, M., Terry, A., Thayer, N., Westerholm-Parvinen, A., Schoch, C. L., Yao, J., Barabote, R., Nelson, M. A., Detter, C., Bruce, D., Kuske, C. R., Xie, G., Richardson, P., Rokhsar, D. S., Lucas, S. M., Rubin, E. M., Dunn-Coleman, N., Ward, M., Brettin, T. S. (2008). Genome sequencing and analysis of the biomass-degrading fungus *Trichoderma reesei* (syn. *Hypocrea jecorina*). *Nat. Biotechnol.* 26(5), 553–560.

Qazi, S. S., Khachatourians, G. G. (2005). Insect pests of Pakistan and their management practices: prospects for the use of entomopathogenic fungi. *Biopest. Int.* 1, 13–24.

Qualhato, T. F., Lopes, F. A. C., Steindorff, A. S., Brandão, R. S., Jesuino, R. S. A., Ulhoa, C. J. (2013). Mycoparasitism studies of *Trichoderma* species against three phytopathogenic fungi: evaluation of antagonism and hydrolytic enzyme production. *Biotechnology Letters* 35(9), 1461–1468.

Rossman, A. Y. (2009). The impact of invasive fungi on agricultural ecosystems in the United States. *Biological Invasions* 11, 97–107.

Saksirirat, W., Chareerak, P., Bunyatrachata, W. (2009). Induced systemic resistance of biocontrol fungus, *Trichoderma* sp. against bacterial and gray leaf spot in tomatoes. *Asian Journal of Food and Agroindustry* 2, S99–S104.

Schirmbock, M., Lorito, M., Wang, Y., Hayes, C. K., Arisan-Atac, I., Scala, F., Harman, G. E., Kubicek, C. P. (1994). Parallel formation and synergism of hydrolytic enzymes and peptaibol antibiotics, molecular mechanisms involved in the antagonistic action of *Trichoderma harzianum* against phytopathogenic fungi. *Applied and Environmental Microbiology* 60(12), 4364–4370.

Sheng, J., An, K., Deng, C., Li, W., Bao, X., Qiu, D. (2006). Cloning a cuticle-degrading serine protease gene with biologic control function from *Beauveria brongniartii* and its expression in *Escherichia coli. Curr. Microbiol.* 53, 124–128.

Small, C. L. N., Bidochka, M. J. (2005). Up-regulation of Pr1, a subtilisin-like protease, during conidiation in the insect pathogen *Metarhizium anisopliae*. *Mycological Research* 109(3), 307–313.

Thines, E., Anke, H., Weber, R. W. (2004). Fungal secondary metabolites as inhibitors of infection-related morphogenesis in phytopathogenic fungi. *Mycol. Res.* 108, 14–25.

Vinale, F., Marra, R., Scala, F., Ghisalberti, E. L., Lorito, M., Sivasithamparam, K. (2006). Major secondary metabolites produced by two commercial *Trichoderma* strains active against different phytopathogens. *Lett. Appl. Microbiol.* 43(2), 143–148.

Woo, S. L., Scala, F., Ruocco, M., Lorito, M. (2006). The molecular biology of the interaction between *Trichoderma* spp., phytopathogenic fungi, and plants. *Phytopathology* 6, 181–185.

CHAPTER 11

MOLECULAR INTERVENTIONS IN FUNGAL STUDIES FOR SUSTAINABLE DEVELOPMENT

CONTENTS

11.1 INTRODUCTION

Presently, the use and success of molecular markers is enormous, and in the near future the explosion in marker development will continue. It is expected that molecular markers will serve as a potential tool to mycologists to evaluate and to manipulate the fungal genome to create as desired and needed by the society. The identification of fungi is a continuous and a complex process. Furthermore, these techniques have just begun to disclose the correct identity of fungal diversity, in that, the absolute majority of fungi still await discovery and formal description. In recent years, genetic polymorphism at the DNA sequence level has provided a large number of marker techniques in different fields of applications. This has, in turn, prompted further consideration for the potential utility of these markers in fungal identification. However, in genetic improvement utilization of marker-based information is depends on the selection of the molecular marker for particular application. The degree of polymorphism, skill or expertise available, possibility of automation, radioisotopes used, reproducibility of the technique and the cost involved in the analysis are the key factors involving in the selection of markers for different application (Alim et al., 2011).

11.2   MYCOTECHNOLOGY

The information obtained from molecular advances could provide an understanding of how the fungal strain can be manipulated by regulating at the biochemical or molecular level with the techniques such as genetic transformation and protein engineering. This will open up wide-range of novel approaches in which the production of valuable products could be improved and reduce or eliminate the production of those with harmful effects. Therefore, maximum exploitation of fungal production system can be expected. Meanwhile, monitoring of the risk assessment of the genetically modified organisms according to the biosafety protocol has to be considered. The content addressed here will give various options that may be employed to improve fungal strains with emerging functional genomics and proteomic studies.

The enormous applications of fungal molecules lead to a great deal of progress in the improvement of the strains for enhanced productivity in recent years. The use of fungi has already been implemented in pigment production, biofuel industry, agrowaste management, pharmaceutical industry, mycoleaching, biofertilizers and pollution control (Gianfreda and Rao, 2004; Mapari et al., 2005; Vicente et al., 2009; Soliman et al., 2013). Day to day increase in the industrialization amplifies the requirements of commercial production of various metabolites such as enzymes, organic acids and antibiotics. The appropriate strain improvement methods for fungal molecules have to be discovered for optimum production. This book discussed on the necessity of strain improvement and the different classical, genetic and molecular methods, which may positively alter industrial applications of fungi. Exposure to the external and internal factors may further alter the characteristics of newly developed strain; hence, these kinds of influences have to be avoided. As there are a lot of fungal molecules yet to be discovered, new methodology for the stimulated productivity is yet to be discovered. The choice of methodology should be so as to make the organism fit for the production at low cost. The specificity in the strain improvement may lower the purification time and cost. Since, most of the fungal products are in relation with spore formation, avoidance of the mycotoxin producing strain has to be done as they hamper sporulation of fungal molecule (Calvo et al., 2002).

11.3 MYCOINFORMATICS

Ten years ago the cost for employing a desirable technology tool was a bottleneck for mycological research. However, currently with the aid of modern molecular and computational techniques we can sequence the entire community of fungi at a cost equal to the sequencing of 1000 specimens a decade ago. The ease to access molecular data has redefined and benefitted all the branches of life science. Indeed, DNA sequence data has provided a valuable source of information relevant to taxonomic classification, evolutionary studies of the organism origin, identification and classification of species in different habitats and their geographical distribution worldwide (Nilsson et al., 2006). Automated processing and interpretation using bioinformatics tools and databases of the DNA sequence data is the only feasible way for the processing of the huge data. The bioinformatics tools facilitates mycologist with a resource for comparative studies of varied range of fungi. The analyzed genomic library documents, comparative study of genomic data, functional annotation and results of large scale analyzes of all the genomes deposited in the database. The phylogenetic identification tools have indeed contributed to the analytical method to the taxonomic assignment of fungal ITS sequence (Nilsson et al., 2011). The bioinformatics tools due to their rapidity in sequence clustering and similarity searches are most preferred in the accomplishment of community level sequencing. The tools represent a competent asset for providing comparative studies with valid clusters of orthologs from fungal genome databases and phylogenetic remodeling by selection of genes with highest informative value at the required taxonomic level in a user defined fungal group. They provide a solution for researchers in pursuit of species identification rather than merely the best possible match scores by automatically separating the identified and incompletely identified sequences. These tools not only provide high-level resolution among fungal community members, simultaneously provide information on fungal taxonomic composition. Needless to say the potentialities provided from these technologies with respect to characterization at community level seem inestimable (Chen and Pachter, 2005; Nilsson et al., 2005).

11.4   CONCLUSION

Fungi have been exploited by humans in many applications for several decades. In all ecosystems, fungi are playing important role in decomposing many pollutants including plant polymers. It has the capability to mineralize various elements and ions. They can also facilitate energy exchange between the above-ground and below-ground systems. However, it is necessary that attempts be made to utilize available fungal species in remediation of environmental pollution. There are many hindrances in the process of mycoremediation. Ecomolecular markers and microanalysis methods can be used to improve the results. From many centuries, fungi are used for many beneficial purposes such as applications of fungi as a biopesticides, as biological control agents, decomposers, biofuel production and in other industries. In near future, new ideas and hypothesis will emerge which will further help in developing the fungi capabilities as beneficial organism. Genetic and proteomic studies are expected to be the main tool for the future development of fungal studies.

KEYWORDS

- **Agrowaste management**
- **Antibiotics**
- **Biofertilizers**
- **Biofuel industry**
- **Biopesticides**
- **Computational techniques**

- **Decomposers**
- **DNA sequence data**
- **Ecomolecular markers**
- **Functional annotation**
- **Functional genomics**
- **Fungal diversity**
- **Fungal ITS sequence**
- **Fungal biomolecules**
- **Genetically modified organisms**
- **Molecular markers**
- **Mycoleaching**
- **Mycotoxin**
- **Organic acids**
- **Pharmaceutical industry**
- **Pigment production**
- **Pollution control**
- **Risk assessment**
- **Sporulation**
- **Strain improvement**

REFERENCES

Alim, M. A., Sun, D. X., Zhang, Y., Faruque, M. O. (2011). Genetic markers and their application in buffalo production. *Journal of Animal and Veterinary Journal* 10(14), 1789–1800.

Calvo, A. M., Wilson, R. A., Bok, J. W., Keller, N. P. (2002). Relationship between secondary metabolism and fungal development. *Microbiol Mol Biol Rev.* 66(3), 447–459.

Chen, K., Pachter, L. (2005). Bioinformatics for whole-genome shotgun sequencing of microbial communities. *PLoS Comput. Biol.* 1(2), e24, doi:10.1371/journal.pcbi.0010024.

Gianfreda, L., Rao, M. A. (2004). Potential of extracellular enzymes in remediation of polluted soils: a review. *Enzyme and Microbial Technology* 35(4), 339–354.

Mapari, S. A. S., Nielsen, K. F., Larsen, T. O., Frisvad, J. C., Meyer, A. S., Thrane, U. (2005). Exploring fungal biodiversity for the production of water-soluble pigments as potential natural food colorants. *Current Opinion in Biotechnology* 16(2), 231–238.

Nilsson, R. H., Kristiansson, E., Ryberg, M., Larsson, K. H. (2015). Approaching the taxonomic affiliation of unidentified sequences in public databases – an example from the mycorrhizal fungi. *BMC Bioinformatics* 6, 178–184.

Nilsson, R. H., Abarenkov, K., Larsson, K. H., Kõljalg, U. (2011). Molecular identification of fungi: rationale, philosophical concerns, and the UNITE database. *The Open Applied Informatics Journal* 5, 81–86.

Nilsson, R. H., Ryberg, M., Kristiansson, E., Abarenkov, K., Larsson, K. H., Kõljalg, U. (2006). Taxonomic reliability of DNA sequences in public sequence databases: a fungal perspective. *PLoS One* 1(1), e59, doi: 10.1371/journal.pone.0000059.

Soliman, S. A., El-Zawahry, Y. A., El-Mougith, A. A. (2013). Fungal biodegradation of agro-industrial waste. *In: Cellulose-Biomass Conversion*, van de Ven, T., Kadla J. (eds.), InTech Publishing, Croatia, pp. 75–100.

Vicente, G., Bautista, L. F., Rodrígueza, R., Gutiérrez, F. J., Sádabaa, I., Ruiz-Vázquez, R. M., Torres-Martínez, S., Garre, V. (2009). Biodiesel production from biomass of an oleaginous fungus. *Biochemical Engineering Journal* 48(1), 22–27.

INDEX